新能源催化转化科学与技术概论

Catalysis as Central Science for Renewable Energy: An Introduction

金 鑫　严文娟　主　编
刘熠斌　张广宇　副主编

中国石化出版社

内容提要

本书全面介绍了太阳能、生物质转化及燃料电池三大可再生能源转化的化学和化学工程原理。太阳能部分介绍了光催化、光电催化和电催化的基本原理,阐述了水分解反应的反应机理、材料设计和工程等内容;生物质转化部分以平台化合物的转化为重点,介绍了制备平台化合物及其下游产物所涉及的主要化学反应及相关催化作用;燃料电池部分主要介绍了电化学、反应堆设计和系统组合等相关技术。此外,本书还提供了关于各类可再生能源使用的能源、经济和环境评估。

本书为英文著作,可以作为高等院校化学工程与工艺、应用化学、环境工程、能源与化工及其他与新能源领域相关专业的教学用书,也可供相关科研与管理工作者阅读参考。

图书在版编目(CIP)数据

新能源催化转化科学与技术概论 = Catalysis as Central Science for Renewable Energy: An Introduction: 英文/金鑫,严文娟主编;刘熠斌,张广宇副主编. —北京:中国石化出版社,2022.6
 ISBN 978-7-5114-6675-4

Ⅰ.①新… Ⅱ.①金… ②严… ③刘… ④张…
Ⅲ.①新能源–催化–英文②新能源–转化–英文
Ⅳ.①TK01

中国版本图书馆 CIP 数据核字(2022)第 089960 号

未经本社书面授权,本书任何部分不得被复制、抄袭,或者以任何形式或任何方式传播。版权所有,侵权必究。

中国石化出版社出版发行
地址:北京市东城区安定门外大街 58 号
邮编:100011 电话:(010)57512500
发行部电话:(010)57512575
http://www.sinopec-press.com
E-mail:press@sinopec.com
河北宝昌佳彩印刷有限公司印刷
全国各地新华书店经销

*

787×1092 毫米 16 开本 16 印张 373 千字
2022 年 7 月第 1 版 2022 年 7 月第 1 次印刷
定价:88.00 元

The Preface

This textbook provides a comprehensive introduction on chemistry and chemical engineering principles for catalytic transformation of renewable energies. The level of presentation is suitable for one semester or even a one-year course of junior or senior students in college. A balanced review and summary of fundamentals has been applied on the content of three key renewable energies, (a) solar energy, (b) biomass conversion and (c) fuel cells. For solar energy, the fundamentals on photocatalysis, photoelectrocatalysis and electrocatalysis have been introduced, with a focus on reaction mechanism, material design and engineering aspects for water splitting reaction. The chapter on biomass conversion is primarily focused on the conversion of platform chemicals. The main chemical reactions involved for preparation of those platform molecules and downstream products and relevant catalysis have been introduced based on of chemical engineering principles. Fuel cell technologies have been introduced with a focus on electrochemistry, reactor design and system combination.

The editors have also noticed that, there is a lack of life cycle view for general chemistry and chemical engineering students. Therefore, this textbook also provides a balanced view on life cycle assessment. For transformation and use of solar, bio- and electro-energy resources, this textbook is aimed to provide an introductory level teaching on the harvesting, processing, upgrading and purification steps. Energy, economic and environmental assessment on the use of various types of renewable energies will be provided in this book. This textbook is published with interesting and diverse on-line learning resources. Please contact Professor Jin (jamesjinxin@upc.edu.cn) for detailed information.

We believe that, this textbook is in the appropriate context for training future chemical engineers for scientific and technological innovation in the area of sustainable chemistry and renewable energy. Professor Xin Jin drafted the textbook while Dr.

Wenjuan Yan collected literature and assisted in compiling the teaching materials, which is being employed as handouts in the course of "Introduction to Alternative Energy and Energy Storage Technology" since 2018. Vice professor Yibin Liu, Professor Jian Shen and Dr. Guangyu Zhang provided important feedback on the content of this textbook. The cover artwork was created by Professor Jin in courtesy of excellent education and academic training by Center for Environmentally Beneficial Catalysis in The University of Kansas. Professor Xin Jin wants to express the deepest appreciation for the guidance from Professor Raghunath V. Chaudhari and Professor Bala Subramaniam. We acknowledge with the greatest appreciation the people for their contribution.

We dedicate this book to the young and talented students in Professor Chaohe Yang's group on Chemical Reaction Engineering, and Jin Research Lab of Nanocatalysis for Energy and Environment in China University of Petroleum.

Raghunath V. Chaudhari
Lawrence, KS, United States
Xin Jin
Tangdao Bay, Qingdao, China

Contents

■ **Chapter 1 Energy Outlook: Fundamentals and Principles** (1)

1.1 Net zero emission in different sectors (1)
1.2 Impact of COVID-19 on net zero emission movement (3)
1.3 Green energy and chemistry (5)
1.4 Chemical reaction efficiency (6)
1.5 Catalysis (8)
1.6 Industrial catalysts (9)
1.7 Important industrial processes and catalysis (14)
1.8 Catalyst deactivation and regeneration (17)
1.9 Selection of reaction media (17)
1.10 Sustainable engineering design (23)

■ **Chapter 2 Solar Fuels and Chemicals** (30)

2.1 Introduction (30)
2.2 Fossil and solar based fuel system (30)
2.3 Mechanism of photocatalysis and photoelectrocatalysis (36)
2.4 Nanostructured catalysts: design principles (62)
2.5 Surface and morphological properties (66)
2.6 Standards for experimental studies (79)
2.7 Overall water splitting (92)

■ **Chapter 3 Biomass Conversion to Fuels and Chemicals** (111)

3.1 Introduction (111)
3.2 Biomass: the ultimate renewable carbon resource (112)
3.3 Fuels from cellulosic biomass (121)
3.4 Lignin for fuels and chemicals (133)
3.5 Platform intermediates for chemicals (152)
3.6 Fuels and chemicals from ocean biomass (167)
3.7 Global movement for biodegradable plastics (167)

■ Chapter 4　Fuel Cells: Fundamentals and Technologies ·················· (173)

　4.1　Introduction ·· (173)
　4.2　History and principles ·· (173)
　4.3　Fuel cell units ·· (176)
　4.4　Fundamentals for fuel cells ·· (186)
　4.5　PEFC ·· (204)
　4.6　AFC ·· (215)
　4.7　PAFC ·· (219)
　4.8　MCFC ·· (222)
　4.9　SOFC ·· (227)
　4.10　Development of modern electrochemistry ·· (233)
　4.11　H_2 as important energy carrier ·· (238)

■ References and Notes ·· (248)

Chapter 01 Energy Outlook: Fundamentals and Principles

1.1 Net zero emission in different sectors

Energy sector contributes to nearly 3/4 of greenhouse gas emissions today. It holds the key to averting the worsened effects of climate change, the greatest ever facing human society. Reducing global **carbon emissions** to net zero by 2050 is consistent with the goal of limiting average global temperature to 1.5 ℃. This goal requires governments to significantly strengthen their policies on energy conversation and developing renewable technologies, then implementing those policies into practices. The number of countries which have pledged to achieve **net zero emission** goal has covered almost 70% global emission of CO_2. Yet most pledges do not underpin the near-term policies and measures.

The path towards net zero carbon emission is narrow. Stay on it requires immediate, effective and massive deployment of numerous clean technologies available. The world economy is estimated to grow by 40% larger than today, but using 7% less energy. More critically, reduction of greenhouse gases is not just limited to CO_2, where CH_4 reduction should reach almost 75% over the coming decade.

Cheaper and greener **renewable energy** technologies give the carbon emission reduction edge to reach the final goal. Electricity from solar and wind mills is the only best way to reduce carbon emission for power generation (Figure 1 − 1). Despite global recession on fossil fuel demand and consumption, annual addition of solar and wind electricity will reach 630 GW and 390 GW by 2030, approximately four-fold higher than in 2020. Solar and wind power provide an essential foundation for this huge transition. Clean electricity is key for future transportation, from current 5% of global car sales to 30% in 2030.

Government supported R&D is essential in accelerating transforming clean energy innovation into demonstration plant and eventual implementation. Critical areas such as electrification, hydrogen, bioenergy and **carbon capture, utilization and storage** (CCUS) today receive only around 1/3 of public funding for relatively more established low-carbon electricity generation units. Consumer choices also play a vital role. For example, purchasing an EV, retrofitting a house with energy efficient technologies or installing a heat pump. Behavioral changes, particularly in advanced economies, such as replacing car trips with walking, cycling or public transport, also provide around 4% of the cumulative emissions reductions (Figure 1 − 2).

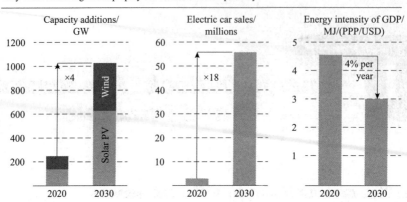

Note: MJ=megajoules; GDP=gross domestic product in purchasing power parity; PPP=purchasing power parity.

Figure 1-1　Capacity increase for electricity and relevant industries

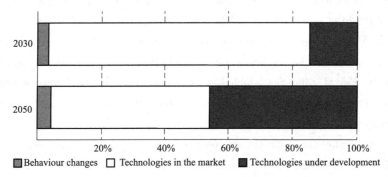

Figure 1-2　Annual CO_2 emission reduction in net-zero carbon strategy relative to 2020

Energy transition also means more opportunities for jobs, leading to larger and deeper social and economic impacts on individuals and communities. It has been estimated that net zero goal will substantiate the employment in energy industry by creating 14 million new jobs by 2030 (Figure 1-3). Further implementation of more efficient appliances, electric and fuel cell vehicles, and building retrofits and energy-efficient construction would require a further 16 million workers.

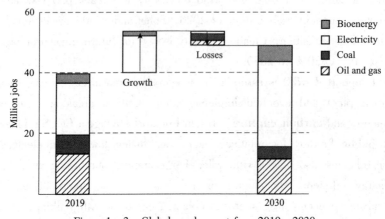

Figure 1-3　Global employment from 2019—2030

Despite severe recession in pandemic 2020, about 110 global companies, which produce large amounts of energy-related or energy-consuming goods have announced net zero emission goals or targets. About 70% of worldwide production of heating/cooling equipment, road vehicles, electricity and cement is from companies which have announced net zero emission targets (Figure 1-4). Almost 60% of gross revenue in technology sector is covered by companies with net zero goals.

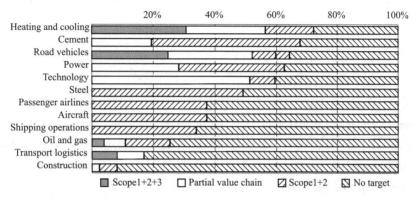

Figure 1-4 Sectoral activity of energy-related companies with announced net zero targets

However, producers from oil and gas sector are largely uncommitted for net zero targets. Only 30% of large energy companies in the world have announced carbon reduction pledges to reach the net zero targets. In addition, the net zero pledge has covered almost 40% of air shipping and 10% of construction sectors.

1.2 Impact of COVID-19 on net zero emission movement

COVID-19 and carbon neutralization are two hot words in the past few years. It has been tragedy years for human beings. Tens of millions of people have been infected or died due to COVID-19. The true number is still rising at the time of writing this book. The combination of the pandemic and actions to be taken to limit the impact of this disease has led to global movement towards a more united community, yet enduring with one of the largest recession in modern world.

Global energy demands have been estimated to have fallen by almost 4.5% in 2020 (Figure 1-5). This is considered as the largest recession since World War II, driven by unprecedented collapse in oil demand and imposition of local/domestic lockdowns decimating transportation demands. Notably, the drop in oil demand accounts for 3/4 of total decline in energy demand globally. Although natural gas displays relatively greater resilience in energy market, it only helped regional market like China to maintain continuing strong growth, less positively impact on a larger scale.

Despite of the chaos and disorders in the past year, renewable energy, led by solar and wind, has been growing continuously and remarkably. Wind and solar capacity has increased by approximately 238 GW in 2020. Accordingly, the share of wind and solar generation in global power mix recorded it largest ever since.

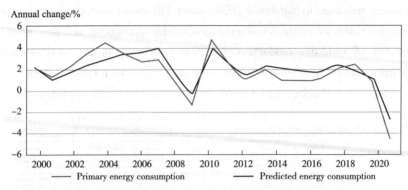

Figure 1-5　Global energy demand: actual and prediction

Since Paris Conference of Parties (COP21) in 2015, there has been a significant number of countries which are striving for **decarbonization**. Approximately 70% of worldwide carbon emission is now covered by carbon neutral or carbon zero targets. Some key numbers for energy development are presented below.

The pandemic has caused a huge economic loss and largest peacetime recession since Great Depression. IMF (International Money Fund) has estimated that around 100 million people have been pushed into poverty as the result of the virus. World energy demand is estimated to have fallen by 4.5% and global carbon emissions from energy use by 6.3% (Figure 1-6). Indeed, overall 2Gt of CO_2 emission has been reduced. The level of worldwide emission has been back as last seen in 2011.

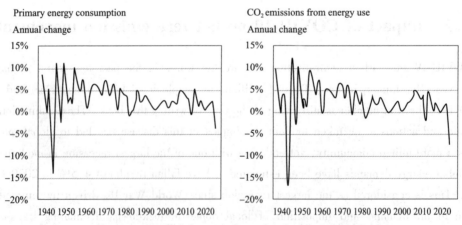

Figure 1-6　Global energy demand and carbon emission

The drop of energy consumption has been driven by oil. Reduction in oil consumption contributes to almost 3/4 of net decline. Natural gas and coal consumption has also decreased significantly. In contrast, wind, solar and hydroelectricity all grew continuously despite of the pandemic. U.S., India and Russia all contributed to the largest decline in energy consumption. However, China actually leads other countries by increasing 2.1% energy consumption with the demands of only a few other countries grew since 2020.

Oil consumption has been reduced by 9.1 million barrel/day (9.3%). Oil demand fell most in

U.S. (2.3 million barrel/day) with EU being 1.5 million barrel/day and India 0.48 million barrel/day. China is the only country where oil consumption actually increased by 220,000 barrel/day.

Price of natural gas has gone through continuous decreasing over the past year, $1.99/mmBTU in 2020 in U.S. and record low LNG price of $4.39/mmBTU in Asia. Natural gas consumption has fallen by 81 billion m^3. But the share of gas in primary energy has risen up to 24.7% in 2020.

Coal consumption has fallen by 6.2 exajoule (EJ, 1EJ = 10^{18} J) or 4.2%, led by huge declines in U.S. (2.1 EJ) and India (1.1 EJ). But China and Malaysia were increasing coal consumption by 0.5 EJ and 0.2 EJ, respectively. Global consumption has been down by 8.3 EJ.

Renewable energy (excluding hydro) has risen by 9.7%, slower than the 10-year average. Solar electricity rose by 1.3 EJ (20%), while wind energy contributes to the largest growth with 1.5 EJ. China was the largest country contributing to renewables growth (1.0 EJ), followed by U.S. (0.4 EJ). Europe contributed 0.7 EJ.

1.3 Green energy and chemistry

As seen in the above sections, energy sectors contribute to large portion of carbon reduction in the future. Development of innovative technologies to reduce energy consumption and carbon emission has been prioritized above all other targets in energy industry. Transformation of primary energy into chemical energy plays a central role. In this context, green chemistry, aiming at providing key principles for sustainable production of fuels and chemicals, has been brought again back to public view.

The classic **green chemistry principles** are being listed in Figure 7. In general, the 12 principles actually cover renewable feedstocks, atomic and energy efficient chemical conversion processes, and avoid use of toxic additives and prevention of waste generation. Those principles have been widely used in developing greener and inherently safer processes. The historical remarks for green chemistry can be dated back to 1969, when former U.S. President Nixon established Citizen's Advisory Committee on Environmental Quality and a Cabinet-level Environmental Quality Council. Environmental Protection Agency was then launched and in 1993, Green Chemistry Program was initiated, to serve as assessment and evaluation of design and processes for their environmental impact. In 1998, 12 principles of green chemistry were published by Professors Paul Anastas and John Warner.

Transformation of alternative resources into chemical energy, namely solar-to-fuel, electricity-to-chemicals/fuels and biomass conversion, involves implementation of manufacture of solar cells, fuel cells, nanostructured catalysts, advanced reactors and energy efficient separation and purification processes. Chemistry plays the central role in all of those processes. For example, manufacture of solar panels, photoelectrochemical catalysts, electrodes, involve application of material chemistry and engineering. In addition, synthesis of **solar fuels**, redox reactions for battery materials, transformation of biomass to fuels all include catalytic chemical reactions.

Figure 1-7 12 principles of green chemistry

Chemical synthesis and transformation using catalyst materials should be environmentally friendly. Catalyst materials include several types such as homogeneous, heterogeneous, biocatalysts and phase transfer catalysts. For energy conversion, in particular for solar, fossil, coal and even electronic energy conversion into chemical energy, catalyst and chemical synthesis is the key, ranging from petroleum refining, synthesis gas upgrading, to starch fermentation, aqueous alcohol reforming.

1.4 Chemical reaction efficiency

Production of inorganic and organic products, ranging from $CaCO_3$, NaCl, H_2SO_4 to alcohols, acids, esters, all demand energy and material input for synthetic processes. The reaction efficiency, involving the ratio of feedstock being transformed and energy intensity is critically important for assessing its environmental impact. This is the basis for production of various industrially relevant products, such as polymers, cosmetics, pharmaceuticals, pesticides, coloring agents, detergents, synthetic fibers and food additives.

In addition, purification and separation steps also demand enormous energy and material management to achieve technological goals. Other potential issues or risks associated with health, safety and social should also be considered and resolved for all processing units.

Applying green chemistry principles and technologies into transformation processes would make production of fuels and chemicals inherently safer and cost-effective. In general, a typical chemical conversion process will generate products and wastes from feedstocks, solvent, reagents and other additives, during pretreatment, conversion and purification steps. Overall generation and discharge of waste will be reduced, if part of the chemicals or wastes can be recycled. Clearly, implementation of green chemistry will undoubtedly reduce materials, energy, risk, hazard and waste.

There is a strong need that quantitative assessment of process/reaction efficiency of chemical/biological synthesis is established. **Yield** is defined as the amount of actual/target product obtained over the theoretical mass of substrates charged or products obtained. **Selectivity** is defined as the ratio of the mole of desired product over that of converted substrate. There are five types of selectivity definitions. Chemoselectivity, diasterioselectivity, enantioselectivity, regioselectivity and stereoselectivity. Measures of effectiveness of chemical reaction are usually named as metric. From the perspective of green chemistry point of view, it is generally accepted that new criteria of environmental compatibility of chemical processes are preferable. E-factor, atom efficiency or atom economy on those important terms.

Atom efficiency is referred as atom selectivity or atom utilization. For example, for a reaction with A and B as reactants and C as the product. Atom efficiency is defined as the ratio of weight of product C over total weight of reactants A and B. Atom efficiency is often used as the important factor to estimate the overall efficiency for multiple-stage processes.

Carbon based efficiency and **mass based efficiency** are used as alternative ways to measure the effectiveness of a multistage process. In particular, carbon efficiency takes into account the amount of carbon in substrates/reactants, which are incorporated into final product. Atom efficiency is very useful, but unfortunately, it does not consider energy input.

$$\text{E-factor} = \frac{\text{Mass of wastes}}{\text{Mass of products}}$$

E-factor was proposed by Roger Sheldon in 1992, which is now accepted as one of the most important measures to assess the environmental friendliness of a chemical process or reaction. This measure adds uncertainty of other associate streams or components during synthesis of targeted products. For example, solvent use and combustion processes may generate extra wastes such as CO_2. The magnitude of E-factor can range from 0.1 to as high as 100 (Table 1-1). For example, E-factor is approximately 0.1 for oil refining sector, while pharmaceutical industry can have an E-factor of 100. It is important to take into account by products and the amounts of waste from each unit, as well as solvent losses, supplementary substance losses, selectivity, refinery waste, and overall yield of a multistep process.

Table 1-1 E-factor for different industrial sectors

Sector	Annual production/t	E-factor
Oil refining	$10^6 \sim 10^8$	0.1
Bulk chemicals	$10^4 \sim 10^6$	$1 \sim 5$
Fine chemicals	$10^2 \sim 10^4$	$5 \sim 50$
Pharmaceuticals	$10 \sim 10^3$	$25 \sim 100$

Mass intensity (MI) is another important measure to quantify the mass ratio of all starting compounds, solvents and auxiliary substances involved in one or multiple stages to the desired products. Mass intensity can be calculated by the following equation:

$$E = MI - 1$$

As a relatively conventional approach, estimating exergy flows for chemical processes is considered as a universal methodology to measure the work function for a system. Exergy is defined as the useful energy which can provide work during a reversible process.

There are several technological synthetic routes for energy molecules. In general, linear synthesis involves gradual transformation of a starting compound. Linear synthesis will include multiple steps and synthetic techniques from upper stream feedstocks to final products. Obviously, fewer stages will benefit the yield. Convergent synthesis can also be utilized where two or more substrates need to be obtained in different sources and transformed into one single target products.

1.5 Catalysis

The concept of catalysis has been widely used in energy and chemical industry. Approximately 95% of industrial reactions involve uses of homogeneous or heterogeneous catalysts. For non-catalytic stoichiometric reaction, turnover number is less than 1 while for catalytic systems, this number is (much) great than unity. Catalytic materials are critically important for catalytic reactions. A **catalyst** is often referred as a substance, which facilitate certain chemical reactions while restraining some unwanted ones, without consuming catalytic material itself. However, during catalytic circle, catalyst itself usually changes, but will return to the original form after the cycle is complete.

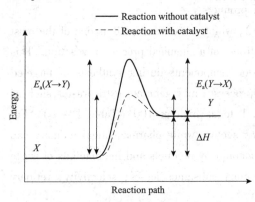

Figure 1-8 Catalysis allow reaction proceed through an alternative path

It is generally known that heat, light, electricity or magnetics are not catalysts, as they only input additional energy source rather than are substances to accelerate chemical reaction rates. The presence of catalysts will alter reaction pathways, therefore, substrates only need to overcome less energy barrier to make this reaction occur (Figure 1-8). One should know that actually the rate (**kinetics**) of a chemical reaction depends on both enthalpy and entropy, as well as concentration of reactants (substrates).

There are so many reactions involving using homogeneous or heterogeneous catalysts. The knowledge of molecular chemistry of catalysis in biochemical (enzymes) and chemical (chemical catalysts) systems share similarities in the action of homogeneous and heterogeneous catalytic materials, which has been the main research focus and key advances in the scientific areas.

Among all types of catalysts, heterogeneous catalytic materials have several advantages. Besides easy separation, heterogeneous catalysts can be used in various mode of operation, such as continuous process, or batch operation. They can be used in fixed bed, slurry bed, fluidized bed, *etc*. Heterogeneous catalysis has been the basics for more than 80% of current bulk chemical products which include both fossil and renewable ones.

Three important laws have been widely accepted in the area of catalysis. Berzelius observed and summarized key points accounting for reaction behaviors. Berzelius believed that catalysts were compounds either in solid or liquid form, have properties affecting the chemical affinity of reactant and promote conversion of the components into other ones/states. Catalysts are not necessarily participating in the reaction and the body does not change. Ostwald further developed the concept of catalysis by defining catalysts in terms of thermodynamic terms: Catalyst does not alter equilibrium between reactants and products. This innovative application of the science of chemical thermodynamics enables selection of the process conditions for reactor experiments to test catalytic materials.

Two basic physical chemical laws are known to govern the thermodynamics and kinetics of catalytic reaction. One is based on van Hoff equation for equilibrium constant K_{eq}.

$$K_{eq} = e^{-\frac{\Delta G^\circ}{RT}}$$

Rate of reaction is affected by pre-exponential factor and activation energy.

$$r_A = Ae^{-\frac{E}{RT}}$$

This approach could be used for evaluating the activity of catalysts. Because it was generally known that, most good catalysts would have lower activation energy, leading to general understanding that, a catalyst will decrease the activation barrier of a reaction.

Sabatier's catalysis law discusses the possible state of reactive intermediates. Under optimal catalytic reaction condition, the chemical intermediates should display medium stability. If they are too stable, they will not be transformed. If the intermediates are too unstable, they are difficult to be activated on catalyst surface.

On the basis of above-mentioned three laws for catalysis, the following mechanism is summarized for catalytic circles.

(a) Molecules need to be adsorbed on catalyst surface for activation. Actually, for heterogeneous catalysis, molecules need to diffuse/transfer from bulk to surface of catalysts/sites for activation.

(b) Molecules undergo rearrangement and recombination, involving cleavage and formation of chemical bond in new molecules, which then desorb from catalyst surface to regenerate catalyst.

1.6 Industrial catalysts

The term **turnover frequency** (TOF), as already mentioned, is an important parameter to characterize the performance of catalysts. It is defined as the amount of reactant being converted over unit catalyst during particular period of time. Accordingly, TON is the total amount of substrate converted per unit catalyst.

The TOF is defined as follow:

$$\text{TOF} = \frac{\text{amount of reactant converted}}{\text{number of active centers} \times \text{time}} = \text{time}^{-1}$$

TOF values may range from 10^{-2} to 10^2 for industrial applications. TON can be calculated by the following equation.

$$\text{TON} = \frac{\text{amount of reactant converted}}{\text{number of active centers} \times \text{time}} = \text{life of catalysts}$$

Typically, TON is in the range of 10^6 to 10^7 for industrial applications. To enhance the productivity of targeted or desired products, selectivity of catalysts towards certain reactions becomes very important, particularly for multiple reactions involved. Several factors should be considered for selection of certain types of catalysts. First of all, the kinetics of reactions need to be fully understood in details. Surface chemistry of catalytic materials will assist design of industrial catalysts. Therefore, understanding the elementary steps for catalytic reactions is critical to evaluate the overall performances.

No matter for homogeneous catalysts in solar irradiation absorption and transport, or for multiphase catalytic reactors for biomass conversion or fuel cell device, catalysts are believed to form complexes with one of the reactants. In the case of homogeneous catalysis, catalysts form intermediate complexes and leave with the formation of a product. For heterogeneous catalysts, several steps could be involved: (a) external diffusion towards catalyst surface, (b) internal diffusion inside catalyst, (c) molecular adsorption of reactants on catalytic active center, (d) surface reaction, (e) desorption of products from catalyst surface, (f) internal diffusion away from catalyst, (g) external diffusion away from catalyst pellet.

Heterogeneous catalysts have advantages such as easy separation and regeneration. But complicated diffusion, adsorption and reaction involved in heterogeneous catalysis may render the overall rate of reactions. For example, slow adsorption of reactants or desorption of products will prevent the regeneration of active centers for further reactions. Still, heterogenization of homogeneous catalysts is still preferable for industrial applications of important reactions.

(1) Homogeneous catalysts

Many industrial catalytic processes require homogeneous catalysts. In particular, oxidation, carbonylation, hydroformylation (Figure 1 - 9), oligomerization, polymerization, hydrocyanation all involve use of homogeneous catalyst materials. Classic reactions such as hydrolysis, Diels-Alder reaction, Cannizzaro reactions all need homogeneous catalysts. Advantages of using homogeneous catalysts include good mixing of reactants with catalysts, promoting remarkable interaction between catalytically active centers with substrates. It could lower the reaction temperature and pressure. Another advantage includes the ease of fundamental understanding on catalytic chemistry and mechanism, because the reaction rate is only dependent on intrinsic kinetics rather than mass transfer rate.

Commonly used industrial homogeneous catalysts include organometallic compounds, mineral acids. Organometallic compounds consist of metal-carbon bond and ligands. Nature of active centers can be tuned by modifying the types of ligand. Transition metals play a key role. This is because the availability of d-electrons at transition metal centers determine the nature of catalysis. But metal complexes usually cannot sustain temperatures higher than 200℃. This has been a limitation for homogeneous catalysts.

Oxidation of cyclohexane is one important case for homogeneous catalysis. Co and Mn complexes or acetates are mixed with reactants in the presence of nitric acid. Oxidation of cyclohexane occurs in homogeneous medium releasing NO_x as by-products (Figure 1 - 10). Rhodium phosphine-based metal complexes such as $[RhCl(PPh_3)_3]$ have been found to be an effective catalyst for the hydrogenation of olefins.

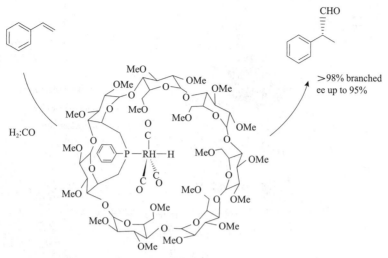

Figure 1-9 Rh complexes for hydroformylation applications

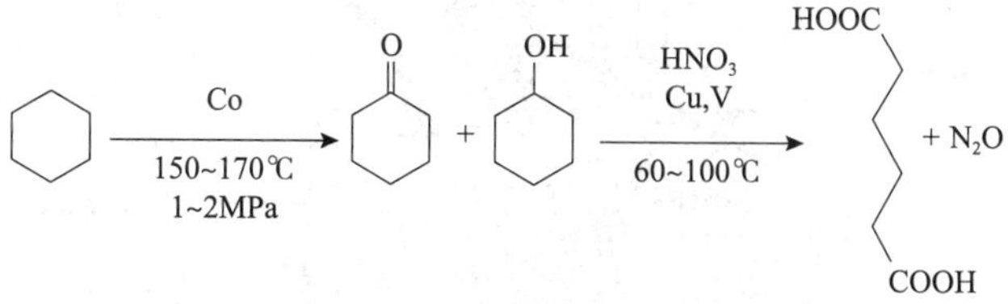

Figure 1-10 Homogeneous Co catalysts for oxidation of cyclohexane to adipic acid

(2) Heterogeneous catalysts

Applications of heterogeneous catalysts, particularly solid catalyst materials can be dated back to 1800s with Faraday using platinum wire to discover oxidation reactions. Numerous types of solid catalysts have been emerging since then. For example, PtRe/Al_2O_3 catalysts for naphtha reforming have been widely used in petrochemical industry. Pt and Cr based catalysts are leading ones for alkane dehydrogenation (Figure 1-11), as developed by commercial technologies. Zeolite catalysts are best known for catalytic cracking of heavy oil into diesel and gasoline products. Synthesis of polyethylene can be achieved in the presence of Ziegler-Natta catalysts [$TiCl_4$-Al(C_2H_5)$_3$]. Solid catalysts are employed for numerous reactions such as hydrogenation, oxidation, nitration, coupling, condensation, hydrolysis, etc.

Solid catalysts have also been widely used in transformation of renewable energies. Solid semiconductor catalytic materials such as TiO_2, Fe_2O_3, metal sulfides are excellent catalysts for H_2 or O_2 evolution reactions. Band gap structures can be modified by solid state and wet chemistry. For biomass conversion, supported metal catalysts such as Pt, Ru, Ni, Cu, etc. display remarkable performances for oxidation of sugars, hydrogenolysis of polyols, hydrogenation of furfurals and dehydrogenation of alcohols. Pt based catalysts are the most important materials for fuel cell devices, where numerous studies have been put to replace expensive noble metals (Figure 1-12).

Catalysis also lays the foundation of solar energy conversion to fuels and chemicals. In this area, both chemical and biological catalysts have been proposed and validated for photon-to-

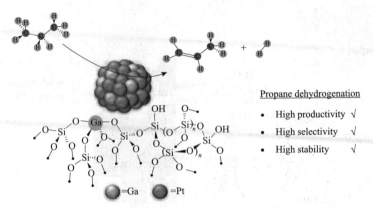

Figure 1 – 11   Pt catalysts for propane dehydrogenation

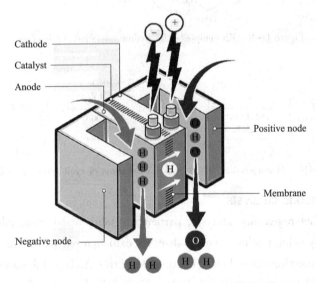

Figure 1 – 12   Configuration of fuel cell unit

electron conversion processes. For water splitting to H_2 and O_2, the endothermic reaction can be accomplished with a minimum energy of 1.23 eV. The electrons excited by photons react with protons to generate H_2, while holes in valence band of semiconductor catalysts oxidize with water to form O_2. A single catalyst can facilitate water splitting reactions upon UV-Vis irradiation (Figure 1 – 13). However, many catalysts suffer from low activity due to incomplete charge separation. Electrons and holes easily recombine to form fluorescence or waste heat. Co-catalysts are often doped to enhance charge separation, lowering the activation or overpotential barriers for water redox reactions. Metals such as Pt, Pd, Ni, Ru can form rectifying Schottky heterojunction trapping electrons to improve H_2 evolution reaction.

Three fundamental limitations still remain to bottleneck the application of H_2 evolution technologies. The efficiency in actual industrial operation is considerably below the thermodynamic limits for the water splitting reaction. The lifetime of electrode materials is too short. There is a lack of a cost-effective replacement for noble metals.

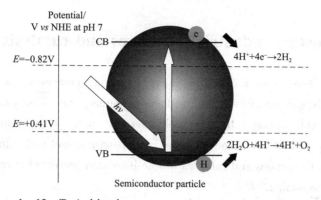

Figure 1 – 13   Typical band gap structure for semiconductor photocatalysts

In Chapter 3, we will discuss the catalytic conversion of biomass to fuels and chemicals (Figure 1 – 14). In particular, lignoncellulosic biomass can be transformed into various fuel additives and valuable chemicals. Lignocellulosic biomass mainly consists of cellulose, hemicellulose and lignin. Cellulose is a polymer of glucose consisting of β-d-glucopyranose units linked by β-glycosidic bonds, accounting for 40%~50% of lignocellulosic biomass. Hemicellulose is a branched polymer with low degrees of polymerisation. Hemicellulose is composed of pentose, hexose and uronic acids, making up approximately 15%~30% of lignocellulose. Lignin, made by aromatic functionalities, is heavily cross-linked with coumaryl, coniferyl and sinapyl alcohols. For example, carbohydrates can be converted into renewable platform chemicals such as 5-hydroxymethylfurfural. Catalysts also play a critical role for this reaction. Various homogeneous and heterogeneous acidic catalysts have been investigated to enhance the chemical yield of 5-hydroxymethylfurfural.

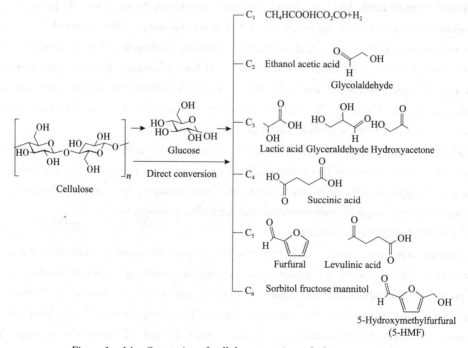

Figure 1 – 14   Conversion of cellulose to various platform compounds

1.7 Important industrial processes and catalysis

Catalysts have been widely used in many industrial processes, such as hydrocarbon conversion, synthesis gas transformation, methane oxidation, *etc*. Catalysts are also critically important for renewable energy. In this book, several energy conversion processes will be discussed, including solar H_2 evolution, biomass conversion and fuel cells. In this section, a few examples on industrial reactions and catalysts will be listed and presented to readers for the benefit of fundamental understanding.

(1) Hydrogenation and dehydrogenation reactions

Ammonia synthesis. It is important for human society. This reaction requires activation of N_2 and H_2 molecules (Figure 1-15). Fe is commonly used as catalyst for strongly binding N atoms and promoting dissociation of N_2 molecules.

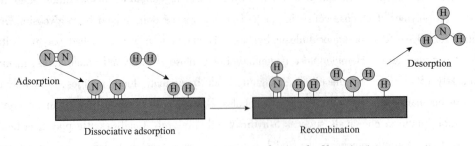

Figure 1-15　Surface reaction mechanism for N_2 activation

Syngas to methanol. Another example is syngas conversion to methanol. In principle, this is a hydrogenation reaction involving reduction of CO into oxygenates. The preferred catalyst in this process is Cu-based because it can dissociate H_2, making hydrogen atoms available to combine with CO. Cu catalysts exhibit limited activity for C—O bond cleavage, therefore, methanol can be formed without further transformation. However, if Cu is substituted by Ni, methane will be formulated instead of methanol. Ni catalysts are relatively more active dissociating C—O bond and adsorbed C atoms will become hydrogenated to methane.

Olefin hydrogenation. Saturation of C=C bond in fatty acid esters is important in food industry. Saturated triglyceride fats with a higher melting point are more useful in the production of margarine. Ni catalysts activate H_2 and bind C=C bond strongly, facilitating reduction reactions rather than C=C cleavage to form methane and light alkane products.

(2) Hydrocarbon conversion

Catalytic cracking. Solid acid catalysts are extensively used as cracking catalysts for transforming heavy oil fraction to lighter ones. Overall the cracking reactions involve converting longer chained hydrocarbons into monomeric or dimeric aromatics, and formation of shorter chain and branched chain molecules. Protonation of the hydrocarbons induces C—C bond cleavage, which occur at elevated temperature due to endothermic feature. Consecutive reactions lead to oligomerization of the olefins and the formation of aromatics. This complicated reaction network, on one hand, leads to the formation of gasoline range molecules, on the other hand, generation of coke with aromatics as precursors. Zeolite catalysts with microporosity are preferable, as it is essential to prevent the formation of larger aromatic rings which are precursors to coke.

Alkane to aromatics. Dehydrogenation of hydrocarbons into olefins is very important, as it provides essential intermediates for numerous downstream chemical processes (Figure 1 - 16). Thermodynamic limitations on dehydrogenation reactions make commercial technologies conducted at high temperatures, leading to side reactions such as excess C—C cleavage and coke formation. Bifunctional catalyst materials are needed, acidic centers for normalization or isomerization, while transition metal sites for dehydrogenation/hydrogenation reactions.

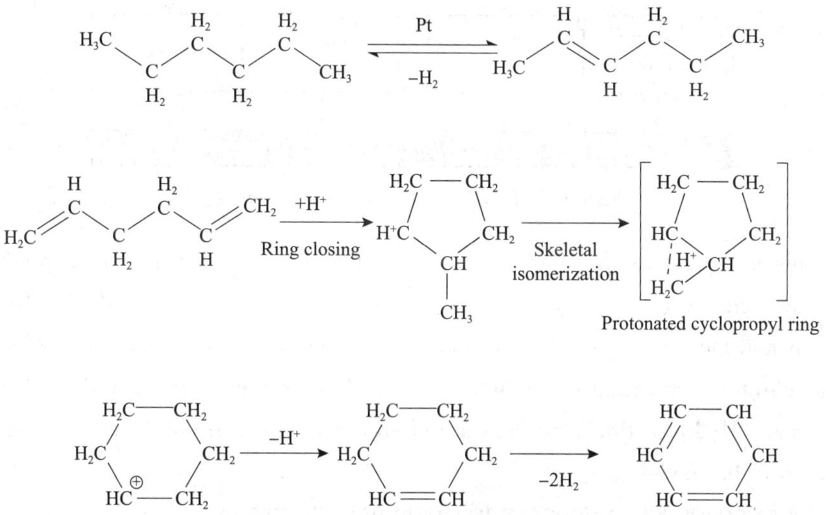

Figure 1 - 16 Schematic description of aromatic formation from n-hexane

(3) Hydrodesulfurization and hydrodenitrogenation

Sulfur and nitrogen content in crude oil and liquefied coal need to be removed for further applications. Similarly, biomass conversion requires selective deoxygenation, desulfurization and denitrogenation to remove O, S and N content. Commercial catalysts for those hydrogenation reactions are MoS_2, WS_2 promoted Co or Ni catalysts. Promoters are added to create surface vacancy position during reaction of H_2S and ammonia. Molecules can be adsorbed on those surface vacancies by C—S and C—N bond activation by transition metal cations. Catalysis consists of a sequence of H_2 additions and C—S, C—N, or C—O bond cleavage reactions. In practical petrochemical processes, those catalysts are generally used in hydrotreating step for crude feedstocks with complex compositions, which may contain deactivating metals (residue from heavy oil) as well as heavy asphaltene molecules.

(4) Bio-refineries

Biomass is considered as the only available organic carbon source on earth. Therefore, use of biomass as feedstocks presents an alternative to produce fuels and chemicals (Figure 1 - 17). In the context of carbon-neutralization or decarbonization, conversion of biomass could potentially contribute positively to reduction of carbon emission, in the total carbon cycle on earth.

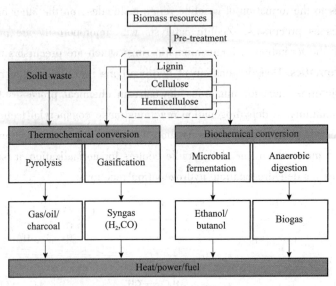

Figure 1 – 17 Biomass conversion to energy

Fast pyrolysis represents the classic way of using bio-energy. This process produces bio-oil with high water content (~28%) and oxygen containing species (~50%). To reduce the oxygen content and meet the standards for liquid fuels, hydrodeoxygenation and hydrotreating can be carried at elevated temperature (<400℃) and H_2 pressure, using sulfide NiMo and CoMo catalyst materials. Hydrotreating processes are often combined with catalytic cracking to optimize the composition of bio-fuels.

(5) Other conversion processes for fuels and chemicals

Steam reforming. This is an important reaction to transfer natural gas, coal, oil and biomass feedstocks into synthesis gas. In particular, those reactants undergo reforming in the presence of water content and generate CO and H_2. Metal catalysts are required, in general, to convert CH_4 in natural gas into CO and H_2. But carbon deposition and formation of inert oxides may deactivate catalysts.

In practice, Ni based solid catalysts are used with wide commercial interests. Alkaline cations or alkaline earth cations are usually incorporated to promote catalyst activity and durability.

Epoxidation. Ag is the most well-known catalysts for epoxidation of ethylene in gas-solid phase reaction. Ethylene epoxide is the intermediate for the production of ethylene glycol for antifreeze and monomer. Selective epoxidation over Ag catalyst is achieved by doping alkali and chlorine species. For propylene epoxidation, the industrial processes are often conducted using Lewis acid catalysts. Propylene epoxide are produced in two types of process, one involves propylene and ethylbenzene into propylene oxide and styrene monomer, while another uses H_2O_2 as the oxidant. Styrene production processes is conducted at 120℃ and alkylation of benzene with ethylene is carried out at 400 + ℃. Propylene oxide is also produced in direct epoxidation in the presence of H_2O_2 as the oxidant. Dow company developed this process using TS-1 zeolite. Such process show a outstanding environmental benefits compared with conventional production processes, as waste water production has been reduced over 80% and energy use lowering by 35%.

NO reduction. Catalytic converters were introduced in 1970s to reduce the content of CO, NO_x and remaining hydrocarbons. Early catalytic converters employed a catalyst consisting of a Pt/Pd/Rh alloy on a reducible oxide such as CeO_2 to reduce the NO content of the exhaust. During reduction of NO, NO molecules first dissociate on noble catalyst surface, where N_2 is formed and the reaction is facilitated by CO to form CO_2.

Exhaust of diesel engines cannot be regulated to eliminate O_2. Urea is often used to react with water to produce NH_3 and CO_2. NH_3 and NO react to form N_2. Catalysts used for this reaction include V_2O_5 and zeolites. Ion exchange with Cu^{2+} and Pt^{2+} leads to formation of more catalytic active sites.

1.8 Catalyst deactivation and regeneration

All catalysts deactivate, sooner or later, at different rates. The rate is dependent on deactivation mechanism. For example, unwanted side reactions, parallel reactions, poisoning of active catalytic sites by impurities or contaminants in the reaction mixture. Or simply by blockage of pores and surfaces owing to the formation of coke during reactions, for example, catalytic cracking. Sometimes, products can bind strongly on catalyst surface thus prolonging overall reaction time. Catalyst structures can also decompose at harsh conditions. For example, catalysts for flue gas treatment can be covered by dust or plugged by fly ash. This is particularly common for power generation plant. Structural decomposition can also occur at high temperatures, in terms of sintering of active metal sites leading to reduction of surface area.

Among various deactivation mechanism, the following are categorized: (a) fouling, coking and carbon deposition, (b) thermal degradation and sintering of catalysts, (c) poisoning of active sites, (d) loss of active phase owing to evaporation, (e) attrition.

1.9 Selection of reaction media

From biomass conversion to photocatalytic water splitting, solvent is very important. Solvents are able to dissolve reactants or products without chemical modifying them. In general solvent molecules are liquid at room temperature and ambient pressure. The production of fuels and chemicals is generally known to consume solvents. Subsequently, solvent is often associated with waste produced during formation of targeted products. Recycling of solvent is not a common practice in chemical industry. According to statistic analysis, only less than 50% solvents have been recycled and reused.

In the context of green chemistry, the choice of solvent lies a critical role in promoting environmental friendliness. For example, it would be better to be nontoxic and relatively nonhazardous, not inflammable or corrosive.

(1) Classification and properties of solvents

Solubility of solute, evaporation rate/boiling point, chemical structure and hazard are commonly used for classification. Good solvents generally require the following basic properties, clear and colorless, volatile without leaving a residue, good long-term resistance to chemicals, neutral reaction, slight or pleasant smell, anhydrous, constant physical properties, low toxicity, biologically degradable, cost-effective.

Tools for solvent selection include the solubility, polarity and catalyst-solvent interaction. Solvent with weak hydrogen bond include hydrocarbons, chlorinated hydrocarbons, nitro compounds. Moderately strong hydrogen bond solvent include ketones, esters, ethers and aniline. Solvents with strong hydrogen bonding mainly include alcohols, carboxylic acids, pyridine, water and glycols.

Apart from performance requirement, some hazardous properties should also be considered, including, inherent toxicity, flammability, explosivity, stratospheric ozone depletion, atmospheric ozone production and global warming potential.

(2) Traditional solvents

Hydrocarbons, chlorinated solvents, oxygenated solvents are commonly used in chemical industry. Hydrocarbons are generally obtained from petroleum fractionation. Benzene was used for degreasing metal prior to 1920s. As its toxicity became more publically aware, other solvents such as toluene was employed to substitute benzene. Kerosene, xylene and other petroleum derivatives are widely used as industrial solvents for cleaning and dissolving purposes. Chlorinated solvents perform better in resins, polymers, rubber, waxes, asphalt and bitumen. All chlorinated hydrocarbons are subject to decompose under the action of light, air, heat, and water. It will produce environmental hazards, particularly to ozone layers. ACS has provided guidance on solvent selection from the perspective of green chemistry (Table 1 - 2).

Table 1 - 2 ACS guide on solvent selection

Preferred	Usable	Unfavorable
Water	Cyclohexane	Pentane
Acetone	Toluene	Hexane
Ethanol	Methylcyclohexane	Diisopropyl ether
2-Propanol	TBME	Diethyl ether
1-Propanol	i-Octane	Dichloromethane
Heptane	Acetonitrile	Dichloroethane
Ethyl Acetate	2-MeTHF	Chloroform
Isopropyl acetate	THF	NMP
Methanol	Xylenes	DMF
MEK	DMSO	Pyridine
1-Butanol	Acetic acid	DMAc
t-Butanol	Ethylene glycol	Dioxane
		Dimethoxyethane

(3) Water as reaction media

Technically, water is a traditional solvent. Particularly, for petroleum and coal refining industry, water is not a commonly favorable specie, as presence of water often induce structural decomposition of zeolite materials and pose hazards to process equipment. However, water is present in large quantities during biomass conversion, H_2 evolution reactions and fuel cell devices. Beyond using no added solvent in a reaction or process, water is probably the greenest alternative we have in nature. Water is highly polar solvent and has a high dielectric constant, contains extensive hydrogen bonding. Critical points of water are also higher than other green solvents such as ethanol, acetone. Non-flammability is probably the most favorable property for water. From an environmental point of view, it is renewable, widely available in suitable quality, and odorless and colorless so that contamination can be easily recognized. Particularly, for sugar processing, hydrolysis reactions occur in dilute aqueous solution. Therefore, presence of large amounts of water is evitable for biomass upgrading. Both fermentation and chemical transformation of bio oxygenates need to take place in the presence of water. H_2 evolution reaction, or water splitting reactions need to occur in the presence of water. Actually, water is both a solvent and a reactant under such circumstance, although electrolytes are also required to enhance electric conductivity. Water is the produce for fuel cell operation, as fuels (e.g., H_2, hydrocarbons) are oxidized by air, where water is a product that should be removed quickly.

Disadvantages of using water as a solvent are also obvious, such as poor solubility for various organic compounds and the moisture-sensitive nature of many catalysts and reagents, which can lead to their deactivation. High heat capacity, in one way would favor temperature maintenance during operation, on the other way, would pose significant challenges during heating and cooling processes, particularly for distillation operation.

But using water in a biphasic system would have several advantages including separating homogeneous catalysts from reaction mixtures and allowing catalyst recycle and reuse to reduce the waste. The ideal process is that, reactants are water soluble and products insoluble in aqueous phase. For example, glucose conversion to 5-hydroxyl methyl furfural (HMF) is conducted in a biphasic system (Figure 1 - 18). Dehydration of glucose occur in aqueous phase in the presence of homogeneous or heterogeneous acidic catalysts, where the as-generated HMF is then extracted into organic phase. Biphasic reactor can facilitate conversion of glucose, while transfer from aqueous to organic phase can avoid decomposition of HMF into side-products.

Hydroformylation, carbonylation, Suzuki-Miyaura and Michael reactions can occur in biphasic systems.

(4) Supercritical fluid as reaction media

Supercritical fluids can be used as alternatives to some of conventional methods for reaction, extraction, fractionation, materials processing, *etc*. Supercritical fluids display many intriguing and unusual properties which make them as great media for separation and spectroscopic studies.

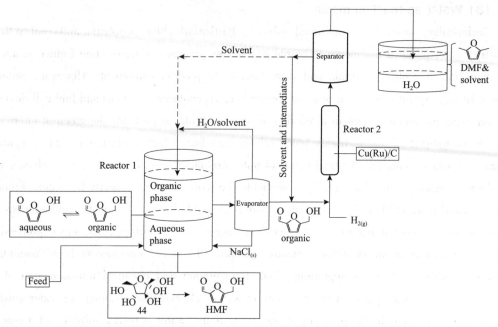

Figure 1-18 Biphasic reactor for conversion of sugars to fuel additives

Supercritical fluids are at state of a compound, mixture above critical pressure and temperature, below the pressure required to condense into a solid. In this region, supercritical fluids exists as both gas and liquid phases. Commonly compounds which could form supercritical fluids include CO_2, water, ethane, ethylene, propane, xenon, ammonia, nitrous oxide. Some substances have readily accessible critical points. For example, T_c for CO_2 is 304K (31℃) and P_c is 7.4MPa, whereas other substances need more extreme conditions. Hence, the amount of energy required to generate supercritical CO_2 is relatively small. T_c for water is 647K (374℃) and P_c is 21.8MPa. The most useful supercritical fluids to green chemists are water and CO_2, which are renewable and nonflammable.

In the view of decarbonization and carbon-zero initiative, using supercritical CO_2, both as solvent and reactive molecules is useful for developing environmentally beneficial technologies, for either fuel production or specialty chemical products. CO_2 is inexpensive and can be obtained as a major by-product from fermentation, combustion. There are a number of advantages using CO_2 in reaction and separation medium. Inert nature makes CO_2 as reliable solvent for numerous applications. From product isolation to dryness, CO_2 can be effectively used by simple evaporation. It is also useful in pharmaceutical synthesis, where no solvent residues can be left when using CO_2 as reactive media. Most processes involving using supercritical CO_2 are mostly conducted at 10~20MPa. In summary, the advantages using CO_2 include no liquid waste, nonflammable, nontoxic, available in high purity (99.9%), low viscosity, gas miscibility, simple isolation, remarkable transport properties, and inert.

(5) Gas expanded liquid

Gas-expanded liquids (GXLs) consist of large amounts of a pressurized compressible gas such as CO_2 dissolved in an organic solvent. CO_2 GXLs allows better oxygen miscibility compared with other organic solvents (Figure 1-19). Solubility of different gases can be tuned therefore, intrinsic kinetic rates can be enhanced accordingly. For example, cyclohexene oxidation can be achieved in the presence of iron porphyrin complexes in CO_2 expanded medium.

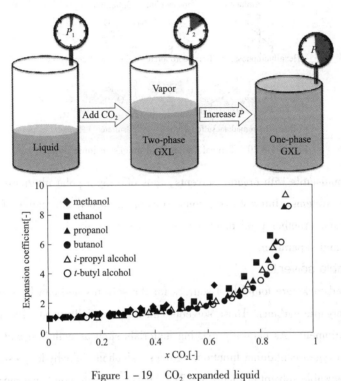

Figure 1-19 CO_2 expanded liquid

CO_2 expanded methanol can be used to extract useful bio-active materials such as lipids and fatty acids from algae. CO_2 expanded liquid can also be applied for heterogeneous catalysis to reduce catalyst sintering and metal leaching.

(6) Ionic liquids

Solvents are often used in bulk quantity, bear a huge cost, and are ranked highly among the damaging chemicals. Volatility of solvents inevitably leads to losses during industrial operation. Therefore, finding stable but promoting solvents is important for developing greener technologies with reduced environmental impact. Typically, an ionic liquid consists of salt where one or both of the ions are large and the cation has a low degree of symmetry. Several other terms such as room-temperature ionic liquids, non-aqueous ionic liquids, molten salts, and fused salts are used as nomenclature for ionic liquids (Figure 1-20). In general, ionic liquids exist in liquid form well below 100℃.

Ionic liquids are widely used as "green" solvents for extractants, electrolytes, sensors, liquid crystals. Inherent low volatility makes ionic liquid advantageous over conventional solvents. Some

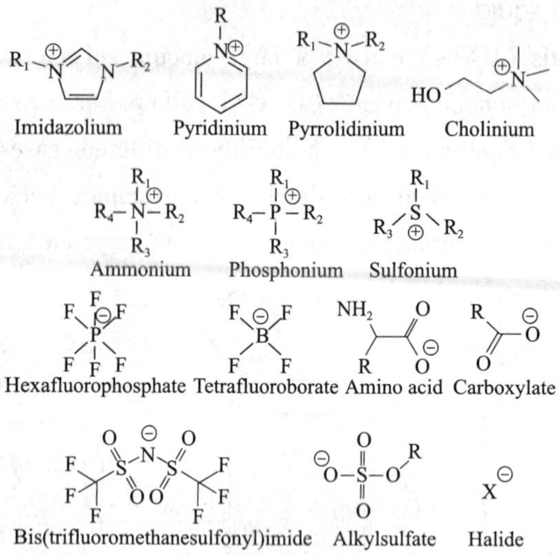

Figure 1-20 Typical cations and anions in ionic liquids

ionic liquids are immiscible with organic solvents, thus offering a polar alternative as non-aqueous nature for biphasic systems. Interestingly, ionic solvents are good alternatives for both inorganic and organic materials. Another good point for ionic solvents is the low temperature operation for catalytic reactions and separation.

(7) Renewable solvents

Renewable feedstocks are found best sources for the solvents and can be used as alternatives to reduce volatile organic carbons. Those solvents can be potentially used without modification in current process equipment. As naturally existing materials such as cellulose and starch consist of large quantities of oxygen-containing functional groups, alcohols, aldehydes, esters and acids are commonly used renewable solvents (Figure 1-21). It is critical to point that mostly used solvents in industries are derived from petroleum, such as chlorinated hydrocarbons. But in view of green chemistry principles, chlorinated hydrocarbons are unfavorable for both human and environment.

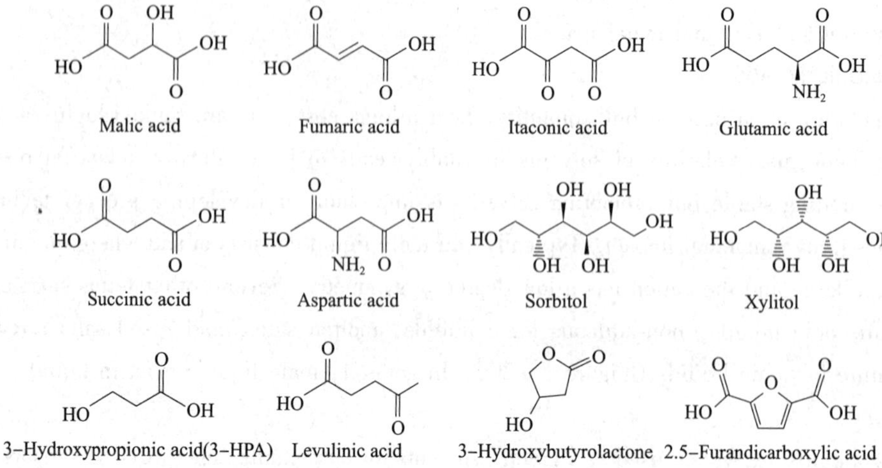

Figure 1-21 Platform chemicals from biomass

Biosourced molecules which are produced either biologically or chemically bear many acids and alcoholic functionalized groups (Figure 1-22). Another way to produce bio-compounds is to convert bio-synthesis gas from gasification and Fischer-Tropsch technology into methanol or hydrocarbons. A variety of bio feedstocks including municipal waste materials, forest products, energy crops, and aquatic biomass materials can be used for production of solvent.

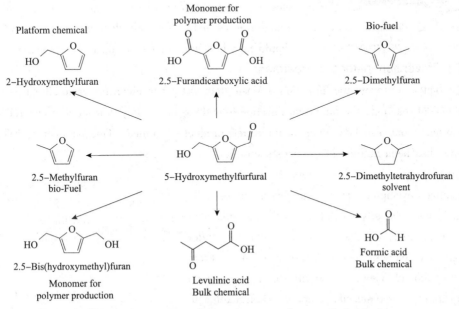

Figure 1-22 HMF as platform for green solvents and chemicals

1.10 Sustainable engineering design

Twelve rules of green chemistry have provided guidance in innovating new chemical reactions and catalyst materials. But applying green chemistry rules into technological development need considerable efforts in transforming laboratory experiments into industrial scales. It needs certain strategy and expertise in scale-up design. The following principles were proposed by Professor Paul Anastas in 2003 formulating the general rules for green engineering.

(1) Inherent rather than circumstantial

The first thing about this principle is that, chemists and engineers should be familiar with the basic properties such as boiling point, melting points, freezing points, vapor pressure, water solubility. Chemical engineers are also more familiar with properties like flammability, explosivity, compressibility, viscosity, and other properties that affect heat and mass transfer.

The second important point for this rule is that, chemists and engineers should investigate new chemical processes from a system perspective (Figure 1-23). They should be able to manage the mass and energy balances around each unit operation, a chemical process, a facility, or an even larger, more comprehensive system including an industrial park or petrochemical complex.

Figure 1-23 Triangle scheme for inherent safe and green process

In addition, process design also considers selecting chemicals or materials made from or with feedstocks/chemicals which will not harm environment. In case of energy input, readily available or easily accessible energy source such as electricity and steam are preferable, although they are obtained from fossil energy. Each type of energy actually poses toxics of certain degree to environment, which should be taken into account in designing processes.

(2) Prevention instead of treatment

It is better to prevent the formation of waste than to treat or clean up waste after it is formed. For a chemical reaction, the important question to address is not only how much of desired product can be formed, but also how much of undesired product is formed. This parameter, referred as selectivity, has been defined in previous sections.

Ethylene epoxidation has often been used as a representative example. This reaction is industrially conducted over solid silver catalysts. Ethylene epoxide, the desired product, can be transformed into CO_2 under harsh reaction temperatures (Figure 1-24). From economic point of view, selectivity towards ethylene epoxide should be maximized, keeping CO_2 formation to a minimum.

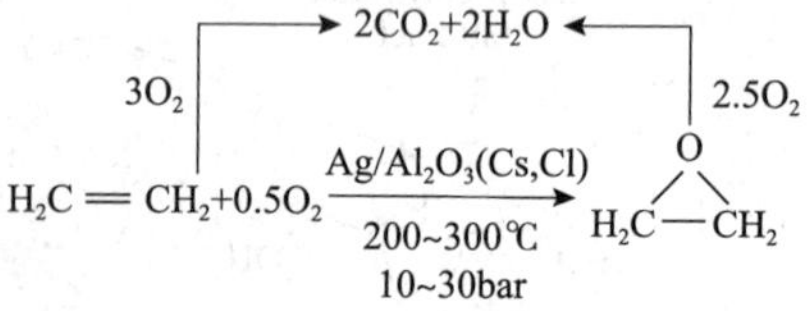

Figure 1-24 Epoxidation of ethylene generates large amounts of CO_2 as by-product

Both technological and operational tools can be employed for engineers and chemists to reduce the amount of waste that is formed in a process. Adjusting reaction temperature can accelerate one reaction over another. If desired products are preferably formed at lower temperatures, lowering temperature may not be able to achieve fast enough production rate to meet economic interest. Concentration of reactants or chemicals also affects the energy consumption during separation process. It is important that the designers include the cost of waste disposal into process optimization.

(3) Design for separation

Separation and purification operations should be designed to minimize energy consumption and materials use. Industrial separation and purification processes can be very energy-intensive, as in most cases, the thermodynamic limits have yet to be achieved. Multiple-stage distillation has been widely used for separating components with different boiling points. From crude oil distillation to bulk chemical separation, significant amounts of CO_2 can be produced owing to large energy input. Therefore, avoiding distillation or making it more efficient is important for green engineering.

Combination of reaction with separation is one approach to integrate complicated processes (Figure 1-25). Reactive distillation is typically used for production of relatively volatile compounds compared with reactants. The key for reactive separation is that, targeted products leave the column in vapor phase, while relatively heavier reactants and other additives in liquid phase.

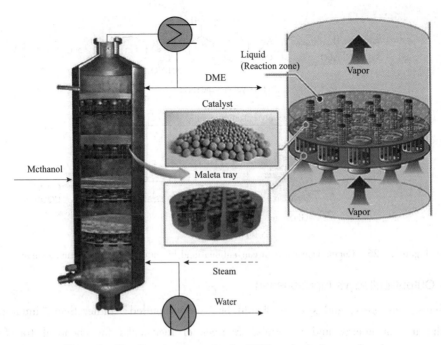

Figure 1-25 Reactive separation for DME synthesis from methanol

Alternative separation technologies, including crystallization, membrane are preferable for lowering overall energy cost. For example, reverse osmosis membrane separation for water desalination is widely applied in water purification.

(4) Maximize efficiency

Products, processes and systems should be designed to maximize mass, energy, space, and time efficiency. Organic chemists are trained to synthesize molecules with desired chemical structures and correct functional groups at right places. This has been done with only the correct final molecular structures, rather than simple and efficient routes. Creative designs should be done to enhance mass, energy, space and time efficiency.

With regard to mass efficiency, it is clear that reactions can be designated to reach as high conversion of substrates as possible, with minimum formation of by-products. A typical example is the production of adipic acid from benzene *vs* from bio-derived substrates (Figure 1-26). Low operating temperature and pressure are desirable, because the need for heating or cooling demands quantities of energy input, thus can be inefficient. In addition, if one can transport minimum mass of materials into next stage, it could be energy saving. This is the result of not needing to pump, stir or temperature control a larger than necessary mass of materials.

In terms of space, or capital investment in real estate, it is obvious that smaller reactor volume will reduce the input for energy and material. The need for expensive construction materials for reactor vessels or processing equipment will also be reduced significantly. For time, short reaction time for completion of reaction means high cost effectiveness. We are identifying opportunities to perform chemical reactions as quickly as possible, while still being as efficient as possible.

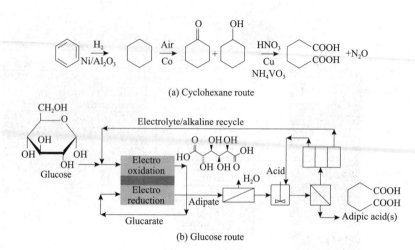

Figure 1-26　Comparison between conventional and bio-adipic acid production scheme

(5) Output-pulled *vs* Input-pushed

Products, processes, and systems should be "output pulled" rather than "input pushed" through the use of energy and materials. Sources of feedstocks for chemical transformation processes are extremely important. The cost, availability and property of input materials will determine the efficiency of processes and overall cost-effectiveness. The importance of inexpensive raw materials to the chemical enterprise cannot be discounted. However, new concept of design was introduced and suggests that the need for product and specific functionality should also be taken into consideration. In this area, 3D printing and electrical cars are important examples. 3D Printing emerges on the call of personalized need for biomedical applications. Electrical cars change the traditional way of power generation and distribution systems. A variety of different products are demanded, from automotive parts to centralized but distributed power generation systems. This is because that convenient use of small-scale converts is badly needed for localized facility such as home and business building. More specifically, solar panels are installed in sunny areas, while wind turbines are installed in windy zones. Excess generation of electric power from solar, wind and biomass can be incorporated into grid and alleviate the need for large-scale power generation plants.

(6) Conserve complexity

Natural products actually evolve through thousands of years to current form. Embedded in this evolution lies a series of network exchanging and interconnectedness, which we have been exploring the reveal the possible tradeoff effect or unintended consequences of human behaviors. As a matter of fact, scientific areas of sustainability and resilience and the contributing fields of green chemistry and green engineering were established in this manner.

One cannot ignore the continuing innovation and advancements in science and technology by applying the principle in one discipline without referring to another. It is in these crossovers where important improvements made in one discipline cannot be lost or minimized when transitioned into another discipline. For example, the innovation or demonstration of improvement achieved at bench-scale chemical reactions are no longer being applied at process-level development. This is because that scale-up issues, such as mass and heat transfer, or mixing effect, are main limiting factors which may restrain the demonstration at large scale.

At another level, attention is needed to address the issue of decomposition or end life of certain products. Favorable features including recyclable and reusable are ultimate designing purposes for either processes and product streams. Such approaches include products made of construction techniques which are easy and fast for disassembly, use of chemicals which have minimal complexity and available in large quantities, molecular structures which allow for facile upgrades without intensive energy and labor input and utilization of minimum amounts of materials.

The logic of conserving complexity is to manage the complexity of the whole system rather than introducing additional energy and material input to make conversion and recycling difficult. In other words, it is important that we think holistically and be mindful of the upstream and downstream considerations of our discoveries.

(7) Durability rather than immortality

Durability of product should be the design goal. However, degradability is also desirable.

(8) Meet need and minimize excess

Need-based design is preferable. Anything excessive over need is not necessary, which might generate side effect, unless for function enhancement.

(9) Minimize material diversity

Introduction of additional components to product or production stream may cause several side effects. For example, formulation of plastic products demand polymerization of monomers, plasticizers, flame retardant and other additives. Materials with multi-components should be minimized to promote facile disassembly and value retention.

(10) Integrate material and energy flows

When one works on chemical synthesis or reaction through a series of synthetic steps at bench-scale, it is easy to focus on the chemistry and atomic efficiency from the perspective of molecular convenience. Even for separation or purification, chromatographic columns and other techniques can easily meet the requirement at lab scale. However, when transferred into a larger scale, a chemical process is a system of interrelated units, inputs, outputs, and recycle streams. Therefore, we need to treat the process through integrating material and energy flow as a whole system (Figure 1-27). Process integration is a powerful tool for system engineering.

Mass and Heat exchange network is important for reducing the overall material and energy cost. Heat exchange network has been widely used by using hot/cold streams to reduce the additional installation of utilities.

(11) Design for commercial "afterlife"

This principle helps us designing new products, or processes, taking into the possible

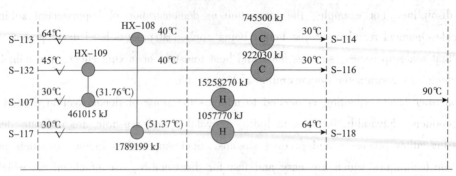

Figure 1-27   Heat exchange network (HEN)

reduction of unnecessary impacts after a product or process has reached the end of its usable life. End-life aspects should be considered, otherwise, literally tonnes of materials will be either landfilled or savaged in unsecure conditions. Recyclable or degradable plastic products are designated to realize this goal. Polyesters should be unzipped to release monomers, which may ben then reused to make new products of other kinds. This methodology can be reused to reclaim the valuables from mixed waste streams, and integrate into polyester manufacture facilities. We can also apply this principle to production or design of electronic products. We can create features which enable recovery of useful materials into higher valuable products.

(12) Renewable rather than depleting

Materials and energies are preferably derived from renewables rather than depleting sources. The fact that the overall population on earth will probably exceeds 10 billion by 2100 is not new. As increased demand meets an ever-shrinking supply of natural resources, prices for commodity materials, everyday essentials and energy will rise inevitably.

On the other hand, environmental pollutions caused by excess use of fossil resources have posed detrimental effect on our natural environment. The amount of marine litter and plastic pollution has been growing dramatically. Emissions of plastic waste into aquatic ecosystems are estimated to triple by 2040, if without any meaningful and effective measures. Actually marine litter and plastics present a serious threat to marine life and even global climates.

Summary

This chapter has introduced the fundamentals on catalysis and energy conversion. Students need to understand the basic concepts for reaction efficiency, green chemistry principles and green engineering rules. Despite of decade development and evolution of those designing rules, one should reflect and apply those principles not only from the view of traditional energy conversion processes, but also alternative or renewable energy/feedstock transformation. Fundamental understanding on catalysis and chemical transformation is critical for learning basic concepts and future design for photocatalytic, photoelectrocatalytic splitting of water, biomass conversion and fuel cell applications.

Exercises for Chapter 1

(1) How energy industry can contribute to reduction of carbon emission?

(2) What is decarbonization? Talk about your understanding on zero-carbon goal.

(3) Please provide examples of industrial processes or laboratory studies using green chemistry rules.

(4) List the advantages and disadvantages of traditional and renewable solvents.

(5) Please conduct a literature survey for a selected industrial process, and estimate the carbon emission for each unit operation.

Chapter 02 Solar Fuels and Chemicals

2.1 Introduction

Solar to chemical transformation provides one of the most promising routes to solve existing energy and environmental issues facing human society. As burning of fossil fuels causes dramatic increase of atmospheric CO_2 on earth. This is the key factor contributing to global warming according to recorded history by global scientists. Chemical conversion of renewable energy to fuels and chemicals, leading to carbon-neutral CO_2 accumulation in the atmosphere, because solar-derived H_2 or carbon-based fuels come from water, CO_2 and carbohydrates. Therefore, in a prolonged period of time, carbon footprint in natural environment is almost negligible.

In this contest, production of solar-derived H_2 and carbon fuels is beneficial in achieving the imperative task of substituting petroleum and coals. This is worthwhile task, as clean production of **solar H_2** fuels provides a sustainable green energy source. For example, H_2 fuels have a remarkably high energy density of 142 MJ/kg.

In this chapter, the fundamentals of solar fuels will be discussed with respect to the most recent advances on rational design of nanocatalysts for efficient H_2 and carbon fuel production from water, CO_2 and renewable recourses. In recent years, the main stream efforts have been devoted to designing innovative nanomaterials for efficient catalytic activation of chemical bond in energy molecules. We will first describe the key role of catalysis in conversion of fossil resources to fuels and chemicals. Renewable electricity, generated from water and biomass is also discussed with respect chemical transformation nature. Photo-induced water splitting to H_2 and O_2 will be then described in terms of mechanism, devices and heterogeneous catalysis. Catalytic CO_2 reduction to fuels and chemicals will also be discussed in details.

2.2 Fossil and solar based fuel system

2.2.1 The energy issues: central of catalysis

It is well established that all fossil fuels were generated by biological catalysis of biomass resources. Biomass, buried underground after death of animals and vegetables, were converted into hydrocarbon-like molecules, through a series of biological and geochemical reaction sequences. These reactions could be carried out in the period of hundred million to billion years and eventually transform biomass into current form of energy molecules. Therefore, in general sense, fossil fuels such as petroleum, gas, coal are not renewable since it is often consumed in short period of time while it takes much longer time to generate in present form (Figure 2 – 1). On the other hand, utilization of fossil fuels in human society also involves complicated chemical reactions promoted by either homogeneous or heterogeneous catalysts. For example, conversion of crude oil often demands zeolite materials as effective catalysts for cracking large molecules into smaller ones, which cover the use of petroleum gas, gasoline, diesel, jet fuels and other intermediate hydrocarbons. The relatively heavier fraction of crude oil undergo further chemical transformation, again, *via* homogeneous and heterogeneous catalytic reactions, into fuels and petrochemicals. Although through different chemistry, coal conversion also includes gasification, hydrogenation, desulfurization processes to obtain fuels and other downstream chemical products.

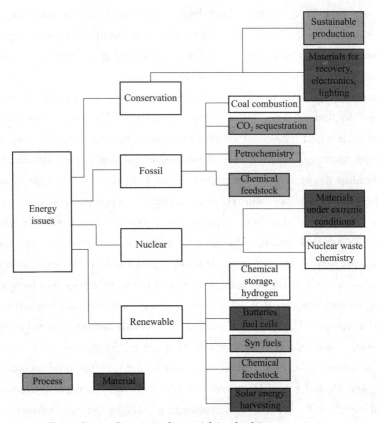

Figure 2 – 1 Process and material involved in energy issues

Undoubtedly, catalytic processes, or catalysis support transformation processes, as well as final liberation of CO_2 from these energy molecules at the end of their lifespan. It is important to point out that, many processes for CO_2 emission are accompanied by energy intensive chemical reactions such as combustion. In the meantime, uncontrollable release of impurities in fossil resources, such as sulfur, nitrogen, as well as other heteroatoms in hydrocarbons, results in serious environmental issues, examples of which range from acidic rain, foggy cities, El Nino effects to ozone depletion, global warming and pollution of aquarium systems. Therefore, modern chemical industry urgently demands clean technologies in converting fossil energy carriers in a controllable fashion, in order to alleviate or solve existing energy and environmental issues.

It is obvious that, chemistry and catalysis stands at the central stage when one comes to seek reliable methods in optimizing the use of fossil fuels. With decades' development, catalysis, reaction chemistry as well as overall system engineering have resolved numerous issues confronting human beings. For examples, R&D on novel zeolitic materials, supported catalyst materials and innovative design of chemical reactors largely lower the energy consumption while enhancing overall atom efficiency for oil refining. Novel technologies have also been applied in CO_2 sequestration in oil field, and nano-scaled materials are developed for CO_2 capture and conversion. Engineers and researchers in industry and academia are devoting extensive efforts in effective use of energy carrying molecules and gaining fundamental understanding of molecular chemistry.

Nevertheless, renewable energy, including solar, wind are biomass are in completely different forms compared with fossil fuels. Especially with rapid development of electrocatalysis and novel nanomaterials in recent decades, people are increasingly aware that, conversion of these renewable energy carriers display emerging and new trends:

(1) If human beings want to address energy and environmental issues in the long run, we should be focused on finding, transforming and storing renewable energy molecules, such as H_2 and bio-derived oxygenates. Chemical Engineering, as a modern scientific subject, established based on chemical reactions involving uses of petroleum and coal, faces significant challenges in 21st century. Because the knowledge we gained in the past 80 + years seem to be incompatible with many of the unique behaviors with H_2 and bio-derived oxygenates, as energy carriers.

(2) There is no doubt that future energy conversion devices would be more like micro-reactors, in a conventional sense. The fast advances in nanoscience and nanotechnologies surprisingly leads the new efforts in finding innovative nanocatalysts for generation of H_2 and bio-derived oxygenates, as well as further transformation of these molecules into fuels and chemicals.

No one would deny the fact that all energy carrying molecules are originally formed due to irradiation of solar energy. Thus, we are putting increasingly attention in using solar energy for synthesizing so-call "solar fuels" in a much faster way, compared with natural processes. Therefore, in the following sections, particular focus is paid on the fundamentals of (a) water splitting into H_2 and O_2 and (b) CO_2 conversion into fuels and chemicals. Shifting the attention from C—C and C—X (X: S, N, O) cracking in existing energy industry, we should be interested in the thermodynamics and kinetics of O—H, C—O and C—H bond cleavage/formation involved in solar fuel synthesis. Similar methodology in investigating the science, but different knowledge would be generated and obtained.

2.2.2 Renewable electricity: the chemistry and catalysis

Combination of the concept of renewable electricity with solar fuels is not new. This idea has been decade's old in terms of combinatory attempts in current energy industry. For example, solar radiation generates wind energy and electricity. Such energy, or energy carriers in a conventional sense, is stored thermomechanically, and then into electricity grid for human consumption (Figure 2-2). Solar fuels, in this figure, particular refers to H_2, generated from water electrolysis or photo-induced splitting, is transformed again into electricity through combustion processes involving turbine devices to generate mechanical energy or electrical energy for **smart grids** as well as our transportation vehicles. Artificial synthesis, in this view, are referred as photochemical conversion of CO_2 and water into hydrocarbons, oxygenates and O_2. The primary products are also considered as solar fuels for smart grid system. The central target besides enabling energy-conserving technologies is to generate chemical energy conversion of electricity.

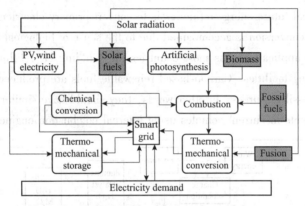

Figure 2-2 Renewable electricity from solar energy

It is of great significance to mention that, the electrical energy is actually stored in the chemical bonds of energy carriers, if one deeply inspects this system. This system also involves an important concept, redox circle. Since energy is stored in chemical bonds, activation of chemical bonds often leads to transformation of chemical energy into electricity. Batteries and storage materials are critical for such reversible process. The battery materials need to respond to electric processes rapidly and properly. In this regard, two key applications, electricity distribution in stationary and mobile units. In particular, the latter one represents the global efforts in both developed and developing countries, on searching powerful and cost-effective battery materials for public transportation system. Actually, the implementation of battery technologies face dual challenges. Massive fundamental research is still underway as many materials are not yet to be affordable to most consumers. On the other hand, we have witnessed that largely empirical science and political efforts dominate current trend have already led to impressive initial successes. This is particular true in China and other Asian countries, where electric vehicles have become indispensables in everyday life.

Back to chemistry, the central role is to enable energy conversion technologies for solar fuels. Therefore, catalysis would be the game-changing factor moving empirical practice to truly scientific and technological breakthroughs in H_2 generation, artificial **photosynthesis** for renewable organic carbon compounds as well as catalytic biomass combustion, all for electricity generation and storage in batteries. It is true that enormous tasking efforts to develop effective water splitting and artificial photosynthesis devices are still at the stage of massive fundamental work. But their impact would be straightforward, as such shift undoubtedly move our daily life from burning of fossil fuels towards extensive use of electricity.

2.2.3 Solar fuels: nanocatalysis

A detailed inspection of solar fuels and solar refineries further reveals the critical role of catalysis in synthesis of solar fuels (Figure 2-3). A new, but controversial trend is clear in academia, and perhaps also in future energy industry, is that we are trying to build a long-standing efforts in establishing a completely different energy system from the traditional one. The central task in the hypothetical future energy refineries is to convert primary electricity and solar heat into H_2. Of course, CO_2 conversion is accompanied due to the sense of biological circle. H_2 fuels have the potential of large applications, as most countries are building H_2 refilling stations with more flexible and cost-saving facilities. Carbon-based renewable fuels are produced to simulate existing ones obtained from fossil resources. In other words, they should have similar molecular structures to meet the requirements in current vehicles using internal combustion engines.

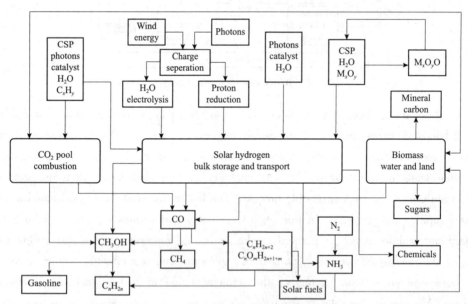

Figure 2-3　Solar H_2 as central molecule in solar fuels

The carbon and H_2 circles are both critical in such hypothetical future energy system. They are both sustainable supplies for energy carriers. Biomass generated via biological processes contributing a major part for future CO_2 capture. But stationary devices are badly needed to collect more CO_2 for solar fuel synthesis. This chapter is dealing with H_2 generation, because it is the ultimate source for transportation tools and essential building block for chemical industry. For example, ammonium synthesis is largely dependent on H_2 industry. The alternative way of producing H_2 will alter the landscape of current solar fuel scheme, which is built solely upon carbon-based molecules, such as CO, CH_4, C_xH_y, etc.

Current opinions on H_2 generation suggest that, we have plenty of room in improving the efficiency of catalytic materials and system hybridization. It is clear that, even up to today, the fundamental challenge on effective water splitting is still unsolved, in terms of efficiency and costs.

Water splitting includes two **half reactions**, one for H_2 generation while the other producing molecular O_2. Water splitting under acidic conditions involves two half reactions, **reduction of protons** and **oxidation of water**. In alkaline medium, H_2 generation reaction occurs through **reduction of water**, while O_2 is produced *via* **oxidation of hydroxide ions** (Table 2-1).

Table 2-1 Half reactions for water splitting

Electrolyte	Half reactions
Acidic medium	$4H^+ + 4e^- \Longrightarrow 2H_2$
	$2H_2O \Longrightarrow O_2 + 4H^+ + 4e^-$
Alkaline medium	$4H_2O + 4e^- \Longrightarrow 2H_2 + 4OH^-$
	$4OH^- \Longrightarrow O_2 + 2H_2O + 4e^-$

Many creative approaches have been proposed to effectively break O—H bond in water molecules in either acidic or alkaline medium. The central focus is the nano-structured catalytic materials. And possible confinement effects on electronic configuration at nano or **sub-nano** scales (Figure 2-4). Water splitting is a reaction where electron transfer occurs as key elementary steps. Rational control of electronic reconfiguration of nanomaterials is the key task in the field of solar fuels, which is essential for future incorporation of renewable H_2 into existing energy grids.

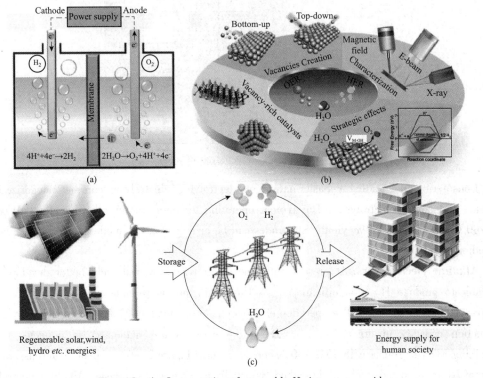

Figure 2-4 Incorporation of renewable H_2 into energy grid

2.3 Mechanism of photocatalysis and photoelectrocatalysis

2.3.1 Fundamentals on H_2 production

(1) Water electrolysis

Typically, electrolysis process consists of an electricity source, which could be obtained from fossil fuel burning or **photovoltaic** processes, electrodes (anode and cathode) and conductive electrolyte. Oxidation reaction occurs at anode and reduction reaction simultaneously takes place on cathode. From economic point of view, large amounts of thermal energy required to split water molecules hinders application of water electrolysis processes (Figure 2-5). Because overall energy is not balanced due to energy loss in converting primary energy sources to electricity. This is the reason that photovoltaics make more energetic and economic sense, owing to the fact that cost of photovoltaic electricity is decreasing fast with improvement in conversion efficiency of solar cells.

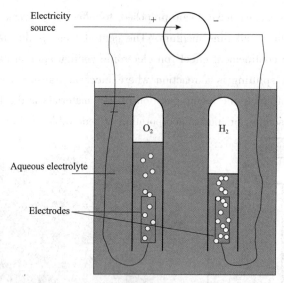

Figure 2-5 Electrolysis of water

Considering the acidic or basic nature of electrolyte, electrolysis can be categorized into alkaline and acidic electrolysis. The typical operating temperature for alkaline **electrolyzers** is 70~80℃, with 70%~80% yield. Electrodes can be arranged in two modes, one as unipolar and the other as bipolar.

Alkaline electrolyzers. At present, alkaline electrolyzers are relatively better developed and available to produce H_2 at significant rates. The equipment is reliable and secure with overall lifespan as long as 30 years. The operation efficiency is approximately 62%~82% with a remarkable production capacity up to 760Nm^3/h. The two electrodes are immersed an aqueous solution containing KOH or NaOH with 25%~30% concentration (Figure 2-6). Anode materials are often made of Ni or Ni-coated steel, while cathode is usually covered with catalysts. Alkaline electrolyzers can be operated at either low pressures (<0.6MPa) or highpressure (~3MPa). The advantages of high pressure operation is that post-compression of as-produced H_2 is not needed in downstream scheme. But decreased purity due to enhanced permeability of membrane is an issue under this circumstance. Economic estimation indicates that the specific energy demands for alkaline electrolyzers is ranging from 4.1 to 7 kWh/Nm^3 H_2 according to operating pressure.

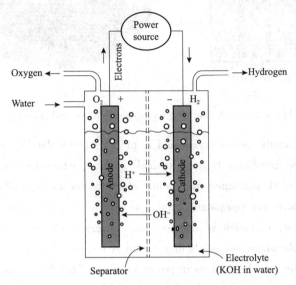

Figure 2-6 Electrocatalytic water splitting in alkaline medium

Efficiency and costs are two determining factors for commercialization of alkaline electrolyzers. Investment costs are ranged from \$1000/kW to \$3000/kW according to recent reports. The following improvements might be considered for future alkaline electrolyzers.

(a) Reducing the size of cells by minimizing the space between electrodes and allowing for higher operating currents.

(b) Developing novel ionic-exchange membranes such as polyacid impregnated with polymers and other sulfoned metal oxide materials.

(c) Enhancing conductivity of electrolyze at elevated temperature above 150℃, in order to promote the kinetics of electrochemical reactions on the surface electrodes. Such operation is favorable for larger scale production of H_2 with up to 99.9% purity.

(d) Due to sluggish kinetics of water oxidation reactions, more effective catalytic materials need to be developed to increase the overall reaction rates.

Compared with acidic medium electrolyzers, alkaline devices are already mature with several commercialized technologies from several companies.

Acidic electrolyzers. Commonly known as proton exchange membrane technologies, acidic electrolyzers are attractive because of the fact that they do not need liquid electrolyzes in the cells. Solid polymer electrolytes are often used allowing a close proximity of the electrodes. Popular membranes include Nafion, commercialized by DuPont, are approximately 0.2 nm in thickness (Figure 2-7). The electrodes are composed of noble metal alloys such as Pt and Ir.

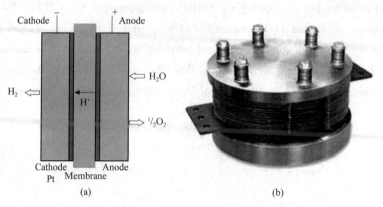

Figure 2-7 Acidic electrolyzer: configuration and assembly

On the surface of anode, water is oxidized to produce molecular O_2, protons and electrons, which travel through the membrane to reach cathode surface, where they are transformed into H_2 molecules. The purity of H_2 generated from acidic electrolyzers are typically 99.99%. Operating temperatures and pressures are comparably similar to alkaline counterparts, with 80% efficiency.

As discussed above, corrosion is a major issue plaguing alkaline electrolyzers. Therefore, acidic cells are favorable where use of alkaline liquid is not possible or inconvenient. The acidic cells are not very sensitive to fluctuations in power supplies. Therefore, many efforts are proposed for potential suitable uses for energy molecules derived from renewable resources. Unlike alkaline electrolyzers, the ion conductivity is not compromised by transporting ions in liquid medium. The main problems associated with acidic electrolyzers are relatively high costs for noble electrodes and membranes.

Other electrolyzers. Different from acidic and alkaline electrolyzers, solid oxide electrolyzers are operated at extremely high temperature (~1200℃). Greater efficiencies can be obtained. This type of device is often composed of a solid metal oxide working in a reverse mode.

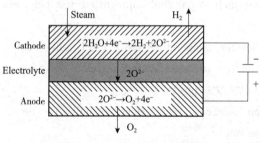

Figure 2-8 Solid oxide cell

The water vapor is pumped into the cathode at elevated temperature to produce H_2 (Figure 2-8). The oxide anions are generated and transferred through the solid electrolyte to the surface of anode, where molecular O_2 is generated. Cathode materials are usually mixed metal oxides such as Ni-based Y and Zr oxide species. Perovskite materials can be used as electrolytes. Optimized design of anode and cathode materials could potentially reduce the energy supply up to 28% by using the heat generated at elevated temperature. The energy sources for this kind of electrolyzer are nuclear reactors, **geothermal energy**, and solar thermal energy.

(2) Solar H_2 production methods

H_2 production using solar energy is classified intro four main categories: (a) photovoltaic, (b) thermal energy, (c) photo-electrolysis and (d) bio-photolysis (Figure 2-9). In fact, thermal energy used for solar H_2 production are usually carried out in two ways, one at low temperature while the other at elevated temperature with concentrated solar energy. Solar thermolysis, thermochemical cycles, gasification, reforming, and cracking are typical high-temperature applications of concentrated solar thermal energy. The other three approaches are considered as mild temperature applications. While bio-photolysis is often referred as biological conversion routes involving metabolism engineering approaches, photovoltaic and photo-electrolysis are receiving particular favors from chemists and engineers in the past decade.

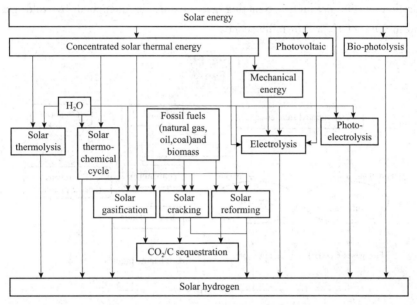

Figure 2-9　Solar H_2 production scheme

Photovoltaic electrolysis. Photo-induced electricity generation has been used for H_2 production since 1970s. The schematic diagram for such process is described in Figure 2-10. Water electrolysis is conducted in the cell where photovoltaic panels are located, to produce H_2 and O_2. The advantage of photovoltaic electrolysis is that, H_2 and O_2 are generated without generating additional greenhouse gases during energy conversion processes. Physical and chemical researchers are extensively working on improving efficiency of photoconverting processes and electrolysis reactions. The current efficiencies are 20% and 80% respectively.

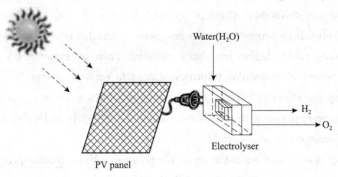

Figure 2-10　Photovoltaic electrolysis

Concentrated solar thermo H_2 production. Direct thermal dissociation of water, known as water thermolysis can give H_2 and O_2 under 2000℃ (Figure 2-11). But the high temperature poses serious problems of materials. Thermolysis of water can be realized in cavity reactors designed for fast heating and quenching functionalities to control the equilibrium of the reversible reaction. When concentrated solar radiations focus on the cavity reactor, they are transmitted through a specially designed glass window to minimize reflection losses. Rapid quenching stops the reversible reaction by reducing the temperature abruptly at a rate of $10^5 \sim 10^6$ K/s and by diluting the concentration of the products. Reacting gas mixture is quenched by cold gas injection, retrieving up to 90% of H_2 within milliseconds.

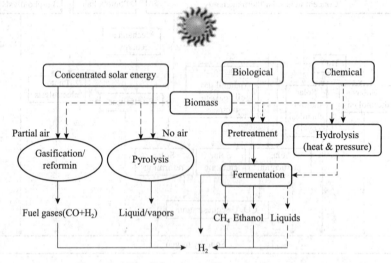

Figure 2-11 Solar thermo H_2 production scheme

It is important to point out that, solar cracking of hydrocarbons and bio-oxygenates also leads to sustainable production of H_2. They are not the focus of this chapter, as we will be primarily paying attention to water splitting chemistry.

2.3.2 Photoelectrochemical water splitting

(1) Solar to chemical energy

Transformation of solar energy is largely dependent on transmitting energy to a confined volume using photonic radiation, which stimulate cleavage of water molecules. Pure water, however, does not absorb radiation in visible and near-ultraviolet ranges. Dissociation of water is technically proceeding under higher-frequency radiation such as extreme UV or X and gamma rays, or in the presence of molecular **photosensitizers** to capture energy in the visible and UV ranges. Considering the effective capture of solar energy within a wider range of wavelength, the interaction of photonic radiation with photo-active catalytic materials and water molecules is critical for technological development in this field.

Water splitting can be carried out at room temperature. Many **photocatalysts** used for water splitting are inorganic materials. The relevant device should show strong chemical robustness and stability. It is impossible to achieve within biological systems. However, the cost must be affordable for sustainable energy conversion using photocatalysis. Water splitting reactions to form H_2 and O_2 is associated with a large positive change in Gibbs energy (237 kJ/mol). H_2 generation from water splitting needs two moles of water to donate four valence electrons in a general reaction according to the following equations

The reaction mechanism is complicated by the four-electron process that occurs in highly corrosive environment. Sunlight can provide essential energy supply for this reaction. But one can see that certain levels of energy needs to be provided in order to enable this reaction by breaking O—H bond. Solar energy should first be absorbed and then transferred into water molecule. Therefore, the energy levels of solar radiation is crucial. Classification of solar energy using solar radiation is shown in Figure 2 - 12. Solar energy can be classified into three categories according to wavelength and energy levels, namely IR (52%), UV (<3%) and visible (44%) regions. IR region is very useful for solar thermal conversion, which has been widely used in civil fields. Visible lights can be utilized for solar photoconversion. The photo induced chemical reactions are extensively recognized in biological and non-biological applications. Photochemical and photocatalytic lie in this category with the aim to better transforming solar energy into chemical energy.

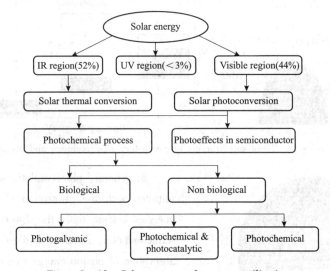

Figure 2 - 12　Solar spectrum for energy utilization

Arguably, the main stream research efforts in both academia and industry are primarily on sophisticated design of nanocomposite materials for better adsorption of solar energy to excite photoelectrons for chemical reactions. Engineering the properties of bulk and interface of nanomaterials must be carried out to satisfy the specific requirements for photocatalysts, or **photoelectrodes**.

The bulk and surface functionalities can be well tuned by varying composition, preparation methods post-treatment conditions of nanocomposites. One of the key issues which researchers are trying resolve is the interfacial properties at solid-solid, solid-liquid. For example, conductivity of electron across the grain boundaries of two distinct catalysts (e. g., semiconductors) is largely determined by the interfacial structures of two components. Electron transfer between solid surface and liquid is governed by the surface and bulk properties of solid electrodes and electrolytes. Obviously, successful implementation of **photoelectrochemical** water splitting at commercial scale requires applications of state-of-art material manufacture technologies.

In general, good candidates for photoelectrochemical water splitting should meet the following specifications: (a) facile excitation of electron-hole pairs in electrodes, (b) charge separation and (c) quick migration of electrons and holes to the surface. In addition, sufficient voltage is needed to facilitate water cleavage reaction.

(2) Mechanism

The chemistry. The theoretical potential for water splitting process is 1.23 eV per molecule. This means that the energy corresponding to wavelength up to 1010 nm, which makes up approximately 70% of solar photons, are capable for driving water cleavage reaction. In other words, the amount of energy provided by photon should exceed 1.23 eV to compensate the intrinsic energy losses during redox reactions on the surface of photocatalysts (Figure 2-13).

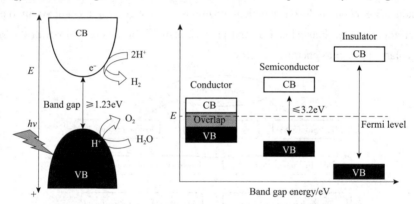

Figure 2-13 Band gap structure of conductor, semiconductor and insulator

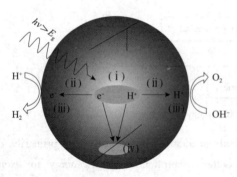

Figure 2-14 Adsorption, excitation, migration and recommbination of electrons and holes

In general, photocatalytic water dissociation is initiated with the absorption of light photons with energies higher than the **band gap** (E_g) of a semiconductor. As described in Figure 2-14, the absorption of photon energy excites photoelectrons in the conduction band (CB) of the semiconductor. Correspondingly, a hole with positive charge is generated in the valance band (VB). Once the photo excited electron-hole pairs are created in the bulk of semiconductors, they tend to migrate to the surface separately. The generation and migration process is competed with electron-hole **recombination** processes. This step leads to the generation of heat in the catalysts.

Under photo illumination in open circuit (no current flow), the excitation of electrons and holes is balanced by instantaneous recombination. It is necessary to mention that such recombination may also occur in an inter-band mode. Recombination in a semiconductor is believed to be pseudo

1st kinetics if photogenerated charges are main carriers but under equilibrium dark concentration, constant carrier concentration in other words. Therefore, it can be seen that, recombination determines the minority carrier lifetime. This is a critical parameter from semiconductor materials. The value ranges from 10^{-3} to 10^{-9} s or even less for many compound semiconductors.

$$L_{min} = \sqrt{D_{min} \tau_{min}} = \sqrt{\frac{k_B T}{q} u_{min} \tau_{min}}$$

Here D_{min} and u_{min} are the diffusion coefficient and mobility of minority carriers, respectively. L_{min} is comparable to the buffer thicknesses on the surface. Surface recombination is very important for photo-energy loss in many devices. This loss contributes to the efficiency loss in light-driven water splitting reaction.

Photon-induced electrons and holes are eventually reacting with water molecules near catalytically active sites in redox reactions. Water molecules are reduced and oxidized by electrons and holes, respectively, to produce H_2 and O_2. With respect to half reactions occurring on catalyst surface, water decomposition reaction actually refers to, either water reduction to H_2 and OH^- species, or water oxidation to O_2 and H^+, or complete water conversion to H_2 and O_2. In other words, water splitting reaction does not necessarily mean a complete water cleavage process. In alkaline medium, H_2 generation occurs through reduction of water molecules with photo-induced electrons, generating OH^- as the co-product. The OH^- species can be oxidized via electro-reaction to generate O_2. In acidic medium, water reacting with holes is oxidized into O_2 and release H^+. Protons are converted to molecular H_2 by combining with electrons generated during photo-excited process.

Catalysts. Clearly, catalyst is the key in improving the photon-induced redox reaction. Catalyst design lies the priority in developing sustainable photoelectrochemical water splitting technologies. Take metal oxide catalysts for example. A CB is similar to the lowest unoccupied molecular orbital, while VB is analogous to the highest occupied molecular orbital. For metal oxides, the electronic structure is distinct from other covalent semiconductors. Such structural uniqueness enable metal oxides to be corrosion resistant and durable in acidic/alkaline medium.

Numerous types of semiconductor materials are suitable for water splitting reactions. Figure 2-15 gives a few examples of semiconductors and present band gap accordingly. ZnS, TiO_2, CdS, CdSe, WO_3 and ZnO are among most popular cost-effective candidates for this application. Commonly studied TiO_2, for instance, has a band gap larger than 3 eV, thus not an intrinsically good candidate for water splitting. Because high frequency solar energy is needed to excite photoelectrons from the molecular orbital in TiO_2 structure. This drawback prevents those materials from actual applications. As most visible-light solar radiation has much lower energy than 3.0 eV. This kind of material is almost insulating. ZnS and ZnO face similar issues. Tremendous efforts have been put to modify the original electronic configuration of abundantly available materials with the aim to reduce the band gap or facilitate generation and reactivity of photo-induced electrons. Common approaches to minimize the band gap include **doping** heterocations, creating oxygen vacancies, incorporation of transition metals and introducing crystalline defects. Some semiconductors have relatively narrower band gap than 3.0 eV, CdS for example. It is highly visible-light active and responsive. But it is not effective for water splitting reaction. Because under illumination process, CdS undergoes degradation in the presence of holes to form Cd^{2+} and S.

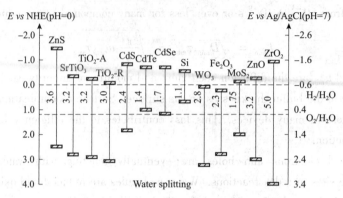

Figure 2-15 Band gap of various semiconductors

(3) Device

In electrolysis of water, an external power supply with approximately 2.0 V is required to overcome 1.23 V energy barrier and potential losses due to polarization. In photo-induced catalytic cleavage of water molecules, external voltage is still required, so called bias voltage (Figure 2-16). For photoelectrocatalytic splitting of water, particle based catalysts and electrode-based devices are both widely used. In particle based system, photocatalysts are dispersed in the aqueous medium. There are both reduction and oxidation reactions occurring on the surface of each particle. Analogously, each particle can be considered as electrodes for water splitting. However, two issues are plaguing with this system. The fast recombination of electrons and holes within the particle causes poor electron-hole separation and migration efficiencies. The as-generated H_2 and O_2 are in mixed form. It poses intrinsic safety issues. But the advantages with such configuration is that the preparation of catalysts and construction of cells are simple and straightforward.

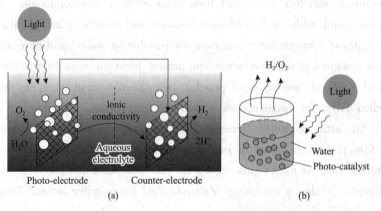

Figure 2-16 Photocatalytic water splitting in (a) electrode-based device and (b) particle-based device

In electrode-based devices, two electrodes are separated and immersed inside the aqueous electrolyte. One electrode consisting of semiconductor catalysts is exposed to light radiation, while the other is a counter-electrode to facilitate electron flow circuit (Figure 2-17).

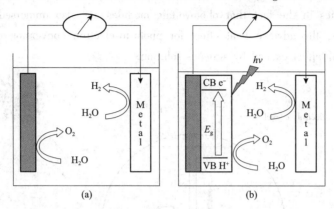

Figure 2-17 Water splitting in (a) electrochemical device and (b) photoelectrochemical device

Other types of photocatalytic reactors have also been studied. Figure 2-18 shows a continuous reactor that utilizes solar light from both sun and a lamp, equipped with electrodes. The advantage of this device is that, sacrificial electron donors (discussed later) can be substituted with the electrode to consume photo-induced holes and transfer charges to metallic sites of the catalysts, which promote H_2 evolution reaction. Solar panels are setup to generate electricity and deliver electrons through electrode-solution interface. UV visible lamps are incorporated inside a sunshine concentrator to collect solar radiation and provide sufficient energy for photoreactions during night time or cloudy weathers.

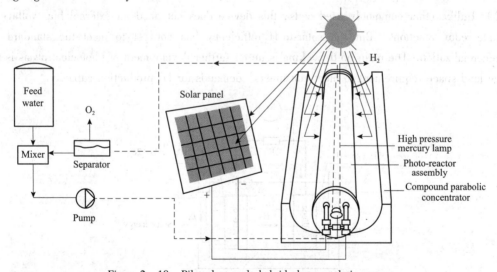

Figure 2-18 Pilot plant-scale hybrid photo-catalytic system

Other photoelectrochemical light harvesting systems were reported by previous researchers. Figure 2-19 presents spherical tank reactor with photovoltaic part for concentrated solar energy conversion device. The Ni cathode and Ni-RuO_2 anodes are immersed inside KOH aqueous solution in a plastic film bag reactor. The reactor body has a function of focusing solar radiation to the surface of electrode. Thus H_2 production rate can be increased accordingly. This reactor is divided into positive and negative terminals by a chamber divider. The structural uniqueness of this reactor also lies in the fact that photovoltaic modules are also immersed inside the KOH solution. Therefore, the adverse heat effect for photo-to-electron conversion can be minimized owing to the cooling effect caused by aqueous solution.

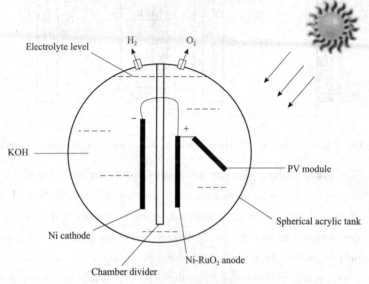

Figure 2-19 Spherical tank reactor with photovoltaic part

The original design by Fujishima and Honda for photoelectrolysis of water is shown in Figure 2-20. Unlike other commonly used cells, this device does not need an external bias voltage to initiate redox reactions. But the solar-to-H_2 efficiency has not yet to meet the standard for commercialization. The main bottlenecking issue for further development of photoelectrolysis is the large land space requirement to meet consumers' demands for H_2 production capacity.

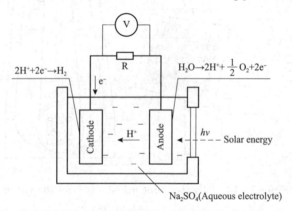

Figure 2-20 Original design by Fujishima and Honda for photoelectrolysis

2.3.3 Surface Reaction Mechanism

(1) Mechanism of photocatalytic cleavage of water

Suppression of electron-hole recombination is one of the key challenges in promoting water splitting efficiency. Introduction of hole scavengers has been employed to enhance electron-hole separation. This process is irreversible as electron donors/hole scavengers are sacrificial agents, however leading to enhanced **quantum efficiency** (Figure 2-21). Therefore, one can visualize the continuous addition of hole scavengers as they are consumed in photoelectrochemical reactions. In general, these agents have lower potentials than electrode materials, thus protecting the surface of electrodes. In this context, sacrificial electrolytes are the main materials of highly interests owing to good capability of controlling electrode reactions.

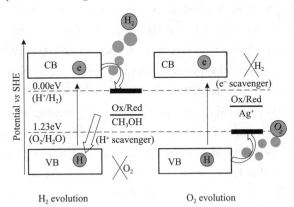

Figure 2-21 Photoexcitation in the presence of sacrificial agents

In particular, a sacrificing agent is added to provide sufficient electron-hole separation as it reacts irreversibly with photogenated holes, minimizing the possibility of recombination while simultaneously avoiding the generation of both H_2 and O_2 in one cell.

(2) Mechanism of heterogeneous catalysis

For heterogeneous catalysis, HER or OER reactions occur at the interface of electrolyte and electrode (Figure 2-22). In general, HER on the catalyst surface includes the most spectacular exposed surface reactions in gas phase and electrochemical systems. There is a direct relationship between the Faradic current (i) and electrochemical reaction rates (dN/dt).

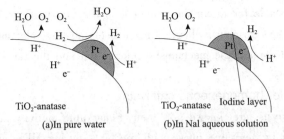

Figure 2-22 Water splitting reactions on the surface Pt/TiO_2 catalyst

$$\text{Rate of electrolysis}\left(\frac{\text{mol}}{\text{s}}\right) = \frac{dN}{dt} = \frac{i}{nF}(\text{homogeneous system})$$

n is the stoichiometric number of electrons which are associated during electrochemical reaction. F is the Faraday constant. Since heterogeneous reactions occur at interfaces, the reaction rate can also be described in the unit of mol/s per unit area.

$$\text{Rate of electrolysis}[\text{mol}/(s \cdot cm^2)] = \frac{i}{nFA} = \frac{j}{nF}(\text{heterogeneous system})$$

j is the current density (A/cm^2). For electron transfer reactions at each electrode, equilibrium potential can be used to characterize the general cathodic charge transfer.

There is no net chemical reaction occurring at equilibrium. In addition, the balanced Faradaic activities can be expressed in terms of the exchange current i_0, with magnitude equal to either of the component current, which is cathodic or anodic current. Similarly, the current density j_0 is the ratio of i_0 to area. The actual potential applied at cathode must be more negative than that posed by equilibrium potential. Reactions can be driven in the presence of activation energy provided electrically. The extra potential, known as **overpotential**, has a key role in governing the efficiency of the electrode-electrolyte system. Activation overpotential, ohmic overpotential and mass transfer effects all contribute to the efficiency of electrochemical reactions.

In fact, heterogeneous electrochemical reactions allow harvesting electrons and holes on the surface of catalytically active sites to ensure chemical transformations occurring at interfaces, eventually leading to the formation of new products. Since HER is critically important for solar energy harvesting for renewable fuels, we discussed here the details of HER elementary step.

$$HA + * \longrightarrow H* + A(discharge\ of\ Volmer\ step)$$
$$H* + HA \longrightarrow e + H_2 + A(atom + ion\ or\ Heyrovsky\ step)$$
$$2H* \longrightarrow H_2(atom + atom\ or\ Tafel\ step)$$

The first step, known as **Volmer step** discharge electrons from H_2 carrier molecules on the surface of catalysts, where H is adsorbed on the active sites for further transformations. Heyrovsky step actually produces more activated H thus enhancing the formation of H_2 molecules. Alternatively, activated (adsorbed) H atoms can be combined to form H_2 molecules as well. Therefore, the overall mechanism involves Volmer step with two possible parallel pathways for H_2 formation.

Validation of proposed mechanistic models can be challenging. Several models describing the elementary steps have been proposed based on measurement at atomic scale in the past few decades. The main obstacle for direct measurement is the fact that, most intermediate species cannot be observed by spectroscopy, classic current-potential-time plot and **Tafel slopes**. Tafel slopes are mostly used to obtain the mechanistic information and the rate-determining step of the overall half-cell reactions.

(3) Mechanism for homogeneous catalysts

Homogeneous catalysts are known for their remarkable activity and selectivity due to homogeneous dispersion in the same phase with reactants during HER and OER applications. Unfortunately, the high solubility and poor durability often hinders the further implementation at larger scale. Homogeneous catalysts can be classified into two categories, organometallic catalysts and acid-base pair catalysts. For overall water splitting reaction, water reduction and oxidation functionalities are needed for homogenous catalysts.

Most of existing studies have been dedicated to water oxidation reactions, as reduction catalysis still remains a grand challenge in this area. The chemical links for water oxidation-reduction catalyst materials are still missing in current studies. The easier structural tailoring can allow more rational synthesis of the simpler mononuclear systems/model systems with respect to the redox-potential leveling, structure-function analyses, and mechanistic studies.

Manganese-oxo complexes. Homogeneous tetramanganese-oxo species are regarded as effective photosynthesis catalysts for O_2 evolution (Figure 2 – 23). Structural arrangement and functionality of the tetramanganese-oxo catalyst are believed to be the key for water splitting. This is a tiny metal oxide cluster composed of four manganese ions, one calcium ion, and five oxygen atoms (Mn_4O_5Ca/Sr), which can take five oxidation states by the redox of Mn ions and stores four redox equivalents.

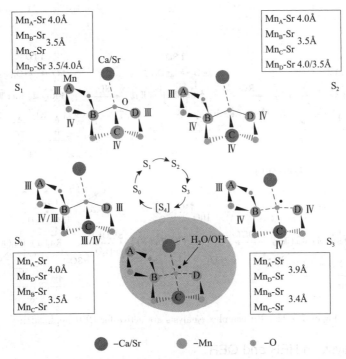

Figure 2 – 23 Mn-Sr catalytic system for O_2 evolution

This figure illustrates the catalytic circle of homogeneous tetramanganese-oxo species for the formation of O—O bond when oxidation states emerge from S_0 to S_4. Water molecules interact with O atoms in the molecular structure and stimulate the formation of O_2 molecules. General hypothesis on the changes of oxidation states is often accompanied with the orbital contribution during redox reactions. In this case, Mn—O moiety has been often observed from *in-situ* studies. The contribution of p orbitals of both oxo groups to a single active site.

Ru complexes. Homogeneous Ru complex catalysts are active for OER application. The low oxidation state Ru^{III}-O-Ru^{III} form of the Ru blue dimer undergoes oxidative activation, when stepwise loss of electrons and protons occurs (Figure 2 – 24). Theoretical studies on Ru complexes revealed the nature of electronic transfer processes. Formation of O—O bond is still considered as the critical step for O_2 evolution. The most plausible mechanism for O—O formation is proposed based on radical coupling of two oxyl radicals, according to spin inversion on the Ru^{III}

center. Computational studies confirmed that, two electrons transfer from Ru^{II} centers to Ru^{III} and give an intermediate species with six unpaired electrons. Inversion of magnetically coupled spin on O^- is facilitated thus two oxyl radicals produce a local singlet diradical par which eventually give O—O bond and peroxide species.

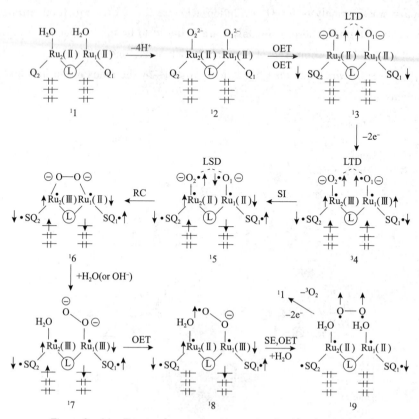

Figure 2-24 Ru complex catalysts are active for OER application

(4) Co-catalysts in HER and OER

For metallic co-catalysts, noble metals are often too expensive for industrial applications. Fundamental studies on non-noble metal co-catalysts are highly desirable in this area. Cu and Ni are among promising candidates. The formation of multiple interfaces in core-shell structures can usually suppress backward reaction for HER and ORE applications. The coating of inexpensive shell can also protect metal core from photocorrosion and improve the robustness of co-catalysts. Metal free OER catalysts have also shown wide attraction. Bioinspired examples such as Co, Mo and Mn containing materials were reported to be efficient for overall water splitting.

In general, many co-catalytic materials are effective for OER applications. Because the four-electron transfer process is referred as the most challenging step. The role of co-catalysts can be illustrated as follow:

(a) Lowing overpotentials for OER which boost the efficiency of photocatalytic water splitting.

(b) Facilitating electron-hole separation at catalyst/co-catalyst interface.

(c) Restraining photocorrosion and improving durability of semiconductor materials.

(d) Providing trapping sites for photo induced charges and promote charge separation.

(e) Suppressing electron-hole recombination and reverse reactions.

Literatures have also demonstrated that appropriate amounts of co-catalysts could be beneficial for photo water splitting, excess presence however, could be detrimental. The possible underlying reasons are stated as follow. The relationship between photocatalytic activity and amounts of co-catalytic materials added is presented in Figure 2 – 25. This volcano shaped curve explain the impact of co-catalyst on the semiconductor catalysts in photo water splitting. Initially, when small amounts of co-catalysts are added to the surface of semiconductors, high photocatalytic activity can be achieved due to the above-mentioned reason. However, excess amounts of co-catalysts could result in blockage of surface active sites of semiconductor materials due to strong coordination interaction, hindering the activation of sacrificing agents or water molecules on catalyst surface. Co-catalysts could also block the irradiation of solar lights on the surface of catalysts. Other important factors such as particle size of co-catalysts, stability and structures also contributes to optimal reaction rates.

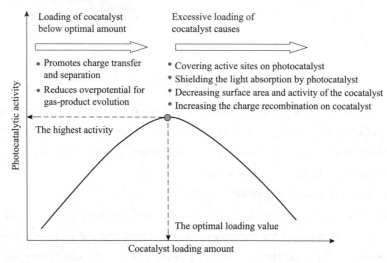

Figure 2 – 25 Influence of cocatalyst loading on photo activity

Hydroxides of non-noble metal materials are often used as effective co-catalysts. $Cu(OH)_2$, $Ni(OH)_2$ and $Co(OH)_2$ loaded on semiconductor materials can improve quantum efficiency (Figure 2 – 26). Such hybridized materials can be prepared using vapor deposition and precipitation methods. The mechanistic illustration for such enhancement can be explained using the following equations.

$$Cu(OH)_2 + 2e^- \longrightarrow Cu + 2OH^-$$
$$Cu + 2e^- + 2H^+ \longrightarrow Cu + H_2$$

The improved performances of photoelectrochemical devices is ascribed to lower potential for $Cu(OH)_2$ to be reduced to Cu and hydroxides, which facilitate HER reactions in alkaline medium by catalyzing reduction of H^+ to H_2 molecules. Another representative case is $Ni(OH)_2$ mediated

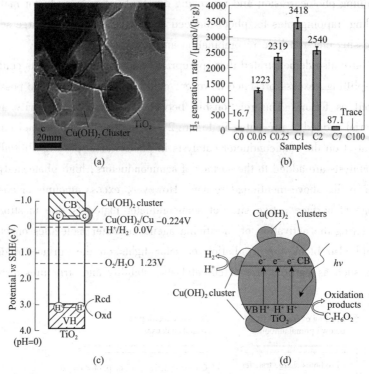

Figure 2-26　Mechanism of Cu(OH)$_2$ enhanced photocatalysis

system. Numerous studies showed that Ni(OH)$_2$ modified metal sulfides, metal oxides, N doped carbons can effectively promote H$_2$ evolution rates. This is because reduction potential is lower than H$^+$/H$_2$. Thus electron transfer from CB to photocatalyst and then Ni(OH)$_2$ can be achieved to favor reduction of Ni^{2+} to Ni^0 and H$_2$ production.

(5) Role of active sites on catalyst surface

Catalytically active sites can exist as an exposed surface plane, edge, kink or defective structures. The nature of actives sites has always been the controversial topic in numerous studies. Existing experimental and computational work has devoted efforts in identifying the properties of materials to reveal the molecular and electronic configurations. As the perspective of identifying active sites has been primarily focused on crystal phases of semiconductors such as TiO$_2$, ZnO, WO$_3$, etc., increasing attention has been paid to the nature of active sites at atomic level. We first analyze the surface energy for selected surface.

$$\gamma = \frac{E_{\text{slab}} - nE_{\text{bulk}}}{2A}$$

E_{slab} is the surface energy, while E_{bulk} is the energy of bulk material, with n as the number of unit cells in the bulk material. A is the surface area of each side of the surface. The stability of the surface under a mild oxidative atmosphere can be calculated as:

$$\gamma_R = \frac{E_{\text{surf}} - \mu_O}{2A}$$

The surface energy of reduced surface can be shown as:

$$E_{\text{surf}} = \frac{E_{\text{Rslab}} - nE_{\text{Bulk}}}{2}$$

μ_O is the chemical potential of O, which is related with O_2 content in atmosphere. These equations can also be plotted in Figure 2-27. The plot suggests that O vacancies start to form on selected surface plane (100, 110) even under O_2 atmosphere. Overall, the ideal surface was found to be stable at higher μ_O, whereas the reduced surface is predominated at lower μ_O.

The electronic structures actually have a determining role in affecting the electron trapping capability. For many semiconductor catalysts, either empty d filled Ti, Zr, Nb and W or fully filled d elements such as Ga, In, Sn, *etc.*, minimizing the electron trapping centers is important as photoelectrochemical reactions require excited electrons and holes moving towards the surface of catalyst materials. Therefore, active sites are usually concentrated on the surface of those metal oxides. Increasing the amounts of active sites by doping another metal oxide often leads to morphological changes. Thus it is clear that, the nature of active sites, considering surface morphologies, energies and electronic configuration affect the performances in water splitting reactions.

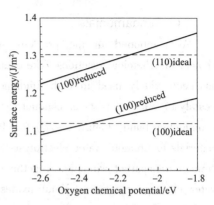

Figure 2-27 Surface energy on oxygen chemical potential

The most acceptable parameter in determining the stability is the **coordination number** of surface atoms (sites). Higher coordination numbers of surface atom closer to the bulk tend to enable more durable surfaces. For example, **rutile and anatase** structures of TiO_2 show varied surface coordination numbers for different facets. Doping of a foreign metal cation could expand the photo response of TiO_2 towards visible spectrum. The chemistry underlying this phenomenon involves induced bandgap shift in the presence of impurity atoms. In CB, dopant ions could be easily photo excited and give hot electrons.

$$M^{n+} + h\nu \longrightarrow M^{(n+1)+} + e^- \text{ (CB)}$$
$$M^{n+} + h\nu \longrightarrow M^{(n-1)+} + H^+ \text{ (VB)}$$

Alternatively, dopant cations could possibly interact strongly with VB and become a hole contributor under solar irradiation. Dopant cations can also lead to electron-hole recombination through a different pathway. The recombination pathway might be favorable for charge separation from original TiO_2 band structures. If the doping is near the surface of semiconductors, it may favor the migration of charges towards the surface. Deeply doped metal cations could behave like a recombination center and make charge migration to surface more difficult. Experimental results show that Cu, Mn and Fe cations can be good electron and hole holding centers, mostly better performed than Cr, Co and Ni ions.

Electron trap: $\quad M^{n+} + e^- \text{ (CB)} \longrightarrow M^{(n-1)+}$

Hole trap: $\quad M^{n+} + H^+ \text{ (VB)} \longrightarrow M^{(n+1)+}$

2.3.4 Heterogeneous catalysis

(1) Fundamentals

As mentioned in the previous section, photon excites electrons in semiconductors for photoelectrochemical reactions. The catalysts are sued in either powder or electrode form. The latter has been widely used in both fundamental research and pilot applications. The photoelectrode (catalyst) often consists of conductive materials, which facilitate transport of electrons from CB (conduction band, Figure 2-28) to VB (valence band), eventually to the active sites of catalytic materials to promote water electrolysis reaction. The photoelectrode is exposed to solar radiation through transparent windows. Together with external bias voltage, catalysts can be activated for water cleavage reactions. Metal oxides and sulfides have been widely considered as potential candidates for future solar-H_2 production catalysts. But some catalyst candidates have relatively larger band gap, thus structural manipulation should be conducted to improve the photo-induced excitation efficiency. Nevertheless, external voltage and coating or doping photo sensitive materials are necessary in almost all cases in order to overcome the potential for water splitting reaction.

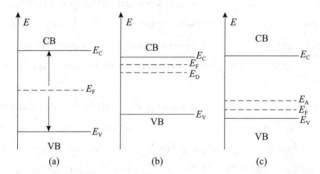

Figure 2-28 CB and VB diagram

Morphologies, size and interfacial properties of those nanocomposites are regarded as key structural parameters for tunable activity for photoelectro redox reactions. Structural defect is also a key factor that governs the behaviors of photo-induced electron excitations within semiconductor structures. For a typical semiconductor catalyst, oxygen vacancy in the bulk phase and surface contributes to upshift of Fermi level energy. Fermi level is an important concept, reflecting the energy of electrons with 50% probability of occupation in orbitals. It can be considered as energy difference between excited free electrons to the ordinary level with half probability of occupation at electronic orbitals. Typically, Fermi level lies in the middle of CB and VB. For n-type semiconductors, where electrons mobility indicates the conductivity, Fermi level is close to CB, whereas for p-type materials, Fermi level is just above VB.

Applying external voltage field also affects the Fermi level of semiconductors. Fermi level can be lifted to relatively higher level in the presence of more negative potentials. Therefore, donors can be ionized even under mild temperature. Excited electrons resides in CB of a semiconductor. Conductivity of electrode are enhanced dramatically owing to sufficient electrons for transport and reactions. This is the reason that most metal oxide semiconductors are known as n-type materials. Electrons in these oxides are main charge carriers, with holes as minor charge carriers.

In general, the distribution of energy levels, which describe the electron occupation probability using Fermi-Dirac function.

$$f(E) = \frac{1}{1 + \exp\left(\dfrac{E - E_F}{k_B T}\right)}$$

E_F is the Fermi energy. When $E > E_F$, the occupation probability falls rapidly towards zero over an energy range of a few k_{BT}, whereas for $E < E_F$, the value rises rapidly to 1. For $E = E_F$, the occupation probability is 0.5. In other words, Fermi level indicates the occupation probability of half in electronic orbitals.

Donor density, or defined as extra number of electrons determines electrostatics at electrode-electrolyte interface. This parameter is measurable. Electrons excited to CB can be thermally or photochemically active for reaction. After excitation of electrons to CB, positively charged holes are generated in VB. Holes can also be move through space as charge carriers.

Apart from oxygen vacancy, doping is another effective approach to increase the total number of extra electrons at electrode-electrolyte interfaces. Doping heterocations into the semiconductor structures alter the original configuration of electronic orbitals. Fermi level will be changed accordingly. Some examples include doping P/As into Si structures. Because the presence of P decrease the energy level of CB thus promoting facile electron excitation for Si materials.

(2) **Interfacial behaviors**

Photon-excited electrons participate reactions at solid-liquid interfaces. To illustrate the process, Figure is presented with details on migration of electrons and surface redox reactions. This process is described as follow:

Following light absorption and activation of active sites on the surface of solid catalysts, certain numbers of electrons are excited to the surface of electrode and migrate to interfaces. With this regard, turnover number (TON) is defined as the number of reactants which are reacted, while turnover frequency indicates the number of converted reactants per unit of time on catalyst surface.

The layer. A layer of ionized solvent molecules, named as **Helmholtz layer** hosts the electron transfer from surface to interface and eventually to the reactant molecules. This layer is adjacent to the solid-liquid, or electrode-electrolyte interface with a thickness less than 0.1 nm. Photo-excited electrons migrate through this layer to reach reactants. To compensate the charge due to loss of electrons, the semiconductor catalyst actually has identical Fermi energy levels as the Helmholtz layer (Figure 2-29). Therefore, electron transfer is iso-energetic. The dislocated electron in CB then falls into a lower energetic level to accomplish the reduction reaction and complete redox cycle.

However, in reality, due to complicated chemical properties of electrolytes and solvents, as well as ill-defined surface morphologies of solid electrode, efforts on characterization of are aimed to reveal the fundamentals to illustrate charge transfer processes and reaction mechanism at interfaces.

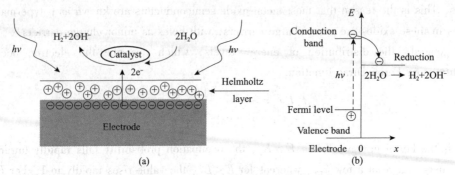

Figure 2-29 Surface of electrode for electron transfer and electron transfer for reduction reaction

Electrode-electrolyte interface. The anisotropic forces at electrode-electrolyte interface and **charge transfer** process occurring at solid-liquid interface leads to rearrangement of electrons/holes inside semiconductors and ions and reactants inside electrolyte. However, the charge transfer processes differ significantly inside semiconductors from metals. The carrier density in semiconductors is much lower ($\sim 10^{18}/cm^3$) compared with metals ($\sim 10^{24}/cm^3$). For metals, sufficient electrons are located near surface (conductor). In contrast, the band gap existing in semiconductors leads to a space charge region of 10~1000 nm distance for charges to migrate to surface of electrode.

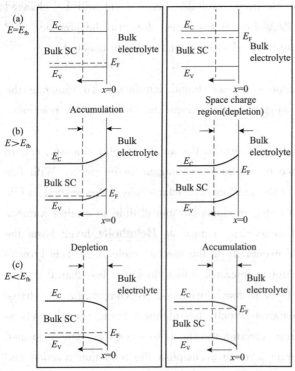

Figure 2-30 Electrode-electrolyte interface

Clearly, the nature of **space charge regions** is determined by the way where charge transfer occurs across the interface. As a result, the band edges in the interior of semi-conductors is dependent with the applied potential similar on Fermi level. Three different situations are to be considered.

(a) If Fermi level of semiconductor is same as the solution redox potential, there is no net transfer of charge. No band bending is found, so-called flat-band potential [Figure 2-30 (a)].

(b) Depletion layer represents a situation when n-type semiconductors show positive of flat band potential. Electrons are depleted due to raised potential of electrolyte at interface. For p-type semiconductors, there decreases the potential of flat band thus charges are depleted in the zone [Figure 2-30 (b)].

(c) For n-type semiconductors, negative potentials of flat-band will lead to excessive accumulation of charge carriers (electrons for n-type materials) in the space, namely as accumulation region. For p-type materials, potentials become more positive than flat-band potential [Figure 2-30 (c)].

Based on the description for **depletion and accumulation regions**, the charge transfer capabilities depends on whether electrons are accumulative or depleted in the thin layer. An accumulative electron region makes electrode more like metallic counterparts. This is because extra electrons/holes in the regions means better charge carrying ability. In the case of depletion layers, few charge carriers are available for charge transfer. But such issue might be resolved by exposing electrode under radiation of sufficient energy. Because electrons can be excited to CB thus more electrons/holes might be available. However, if this happens in the interior of a semiconductor, recombination of electrons and holes induces extensive heat generation. If it occurs at space charge region, the electric field imposed on this region leads to facile separation of charges.

On the electrolyte side, buffer layers can be formed. One is the compact layer (Helmholtz) and diffuse layer (Gouy-Chapman). For depletion region, the positive excess charge formed by immobile ionized donor states, cause electron depletion for n-type materials. As a result, the n-type semiconductor become a p-type at interface, corresponding to an inversion layer. The inversion layer confines electrons mobility due to enhanced mobility of holes.

Band-bending and relevant changes in charge carrying ability can be illustrated using the following example. For an n-type electrode with positive potentials, the band edges shift upwards to allow more holes move towards electrode-electrolyte interface. But electrons are moving towards interface rather than interior phase. Thus electron-hole recombination is not easy to occur. Therefore, n-type materials can act as **photoanodes**. For p-type semiconductors, negative potentials than flat band enable electrons moving towards interfaces while holes are moving to interior part of electrode. Therefore, p-type semiconductors can be used as **photocathodes**.

Figure 2-31 shows an example for n-type semiconductors. No illumination and no electrolyte contact leaves the n-type semiconductor no bending. But when the material is immersed in the electrolyte, band-bending occurs with positive rise of potentials. Accordingly, Fermi level close to CB actually moves down. In other words, electrons are difficult to be excited into CB due to this bending. Under illumination, the bending become less severe because sufficient photon-induced electrons flow towards interior of n-type semiconductors. Additional electrons can be obtained from electrolyte across the interface. Therefore, the photoanode can work properly to continuously transfer electrons from electrolyte to interior of anode. Figure 2-31 further illustrates the setup of a three-electrode system. Solar light can illuminate on photoanode to induce oxidation of water molecules into H^+ and O_2, while electric current is transported through external circuit and H^+ can obtain electrons at photocathode and generate H_2.

The electrode-electrolyte interface enables the formation of **junction** structures (Figure 2-32). The Gouy layer is a diffuse layer close to electrolyte side, while the inner and more compact Helmholtz layer hosts charge separation and transfer across the interface. When concentration of charge carriers increases, the thickness of Gouy layer will decrease. In general, the thickness of Gouy layer is affected by the ionic strength in the aqueous solution. The thickness can be calculated by the following equation:

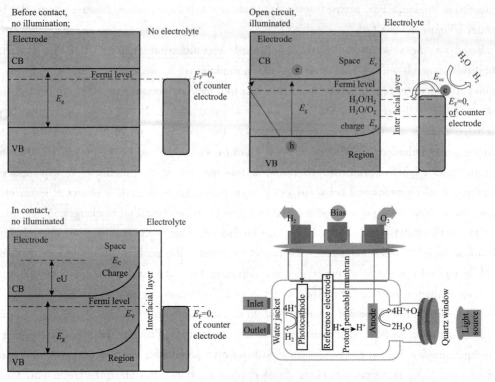

Figure 2 – 31　Changes of Electrolyte-electrode interface under illumination

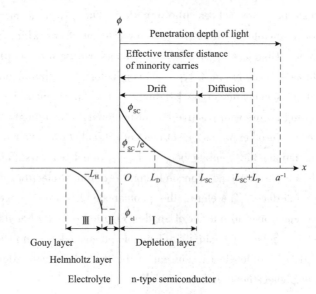

Figure 2 – 32　Composition of semiconductor-electrolyte junction

$$L_{SC} = L_D \left(\frac{2q | \phi_{SC} |}{kT} \right)^{0.5} = L_D \left[\frac{2q(V - V_{fb})}{kT} \right]^{0.5}$$

$$L_D = \left[\frac{\varepsilon_o \varepsilon kT}{q^2 (n_o + p_o)} \right]^{0.5}$$

For photoelectrode, V_{fb}, ε, n_o, and p_o represent the flat potential, the relative dielectric constant, the electron and hole concentration, respectively. L_D is determined by the concentration of charge carriers; ε_0 represents the dielectric constant in vacuum; k is the Boltzmann constant; T is temperature in K.

In addition, the fluctuations of solvent dipoles arising from the thermal energy $k_B T$ may induce temporal variations of electron affinity and ionization energy barriers. Thus the electronic energies can be described by Gaussian probability distributions from the following equation.

$$W(E) = \frac{1}{(4\pi\lambda k_B T)^{0.5}} \exp\left[-\frac{(E-E^\circ)^2}{4\pi k_B T}\right]$$

Here E_0 is the most probable electronic energy level and λ is the reorganization energy. The distributions $W(E)$ shown in Figure 2-33 are the basis of the Marcus theory of electron transfer.

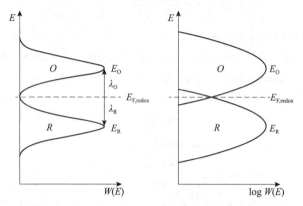

Figure 2-33 Probable electronic energy level

(3) Important parameters

Energy conversion efficiency is very important to access the performances of photoelectrode in solar-to-electricity and electricity-to-H_2 transformation. Efficiency is critical in designing photoelectrochemical reactors for solar H_2 production. Specifically, **incident photon-to-current efficiency** (IPCE) of electrode is one of the key parameters in determining the performances of photoelectrochemical cells. IPCE is influenced by several factors, such as light absorption coefficient, width of space charge layers, charge carrier diffusion paths.

The value of IPCE can be obtained by conducting measurement in either a two-electrode or a three-electrode configuration. Two-electrode system may face big uncertainties due to errors induced by varied properties of **counter-electrode**. Three-electrode configuration is believed to provide reliable data, because it has the fundamental advantage by introducing a reference of counter-electrode. More importantly, a direct comparison with **reversible hydrogen electrode** (RHE) at the same potential can be used to estimating IPCE in three-electrode configuration. We will interpret the measurement accordingly using three-electrode configurations. Typically, IPCE at a predetermined using the following equation.

$$\text{IPCE}(\lambda) = \frac{\text{Total energy of converted electrons}}{\text{Total energy of incident photons}} = \frac{\frac{J_{photo}(\lambda)}{e} \times \frac{hc}{\lambda}}{P(\lambda)} \times 100\%$$

For a fixed wavelength λ (nm), J_{photo} represents the photocurrent density (mA/cm^2); P(λ) represents the incident light intensity (mW/cm^2); e, h, c are the charge of an electron, Planck's constant, and the speed of light, respectively.

It is also worth noted that, optical loss are actually ignored in the above equation. Therefore, to correct the deviation induced by the above-mentioned approximation, we often use **absorbed photon-to-current conversion efficiency** (APCE) to identify the critical role of quantum efficiency. APCE is defined as the total amount of photogenerated charge carriers (e.g., electron or hole) contributing to the photocurrent per absorbed photon. The following equation is presented. A, R, and T represent the absorption, reflection, and transmission of light, respectively.

$$APCE(\lambda) = \frac{IPCE(\lambda)}{A(\lambda)} = \frac{IPCE(\lambda)}{1 - R - T}$$

One may be interested in the overall conversion efficiency, namely solar-to-H_2 conversion efficiency to evaluate the performances of photoelectrochemical reactor. Because it reflects the intrinsic efficiency of H_2 generation under sunlight irradiation (1.5GAM, 100 mW/cm^2) using the following expression. In particular, 273 kJ/mol represents Gibbs free energy of the reaction for water splitting, r_{H_2} represents the rate of H_2 production, P_{sun} represents the incident light intensity (100 mW/cm^2), and S represents the illuminated area of the photoelectrode.

$$\eta_{STH} = \frac{273 kJ/mol \times r_{H_2} mol/s}{P_{sun} \times S \, cm^2}$$

It is the **Faradic efficiency**. Commonly, to meet the energy demand, a water splitting system should show ~10% STH value, and be stable over an extended period of time, to be cost comparable to fossil fuels. The output for visible solar light is much higher (40%) compared with UV spectrum (4%).

$$QE(\%) = \frac{273 kJ/mol \times r_{H_2} mol/s}{P_{sun} \times S \, cm^2}$$

Quantum efficiency can be estimated in two different ways. It can be defined as the fraction of photon input that contributes to photogenerated current. It is estimated as the number of excited electrons generated per incident photon. However, this does not consider the consecutive photons produced by amplification processes. In most cases, the number of photoelectrons generated and emitted is much lower than absorption of photons of light energy. But it could be possible that, due to amplification processes, the value of quantum efficiency might be higher than 100%. Again, this definition neglect the loss of solar irradiation and chemical conversion efficiency. But it reflects the overall performances of a photoelectrochemical cell. Therefore, it can be used to quantify the performances of photoactive catalysts, rather than representing the water splitting efficiency.

$$QE(\%) = \frac{\text{number of evolved } H_2 \text{ molecules} \times 2}{\text{number of incident photons}}$$

This parameter can also be represented using the following equation. Because quantum efficiency is related to term spectral responsivity, therefore, for a given wavelength, one can measure the value of responsivity in a three-electrode system mentioned above. Considering the specific responsivity to a particular wavelength, quantum efficiency has a different unit: amperes per watt (A/W). The relevant equation is shown as follow.

$$QE_\lambda = \frac{R_\lambda}{\lambda} \times \frac{hc}{e} \approx \frac{R_\lambda}{\lambda} \times 1240 \text{W} \cdot \text{nm/A}$$

In this equation, λ is the wavelength in nanometers, h is Planck's constant, c is the speed of light in a vacuum, and e is the elementary charge. R_λ is the responsivity (A/W).

Those well-defined parameters are widely used to estimate the performances of existing photoelectrochemical cells. We should note that the efficiency of solar-to-electricity and electricity-to-H_2 is largely dependent on the properties of photoelectrode, nature of electrolyte, interfacial behaviors and connection of each part in the configuration of electrode system (Figure 2-34). In particular, charge separation, migration and recombination play crucial roles in determining the performances of semiconductor materials. From

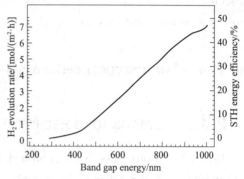

Figure 2-34 Theoretical relation of H_2 evolution rate and solar-to-H_2 efficiency to band gap

materials point of view, the crystallinity and surface functional groups or atomic alignment of active sites influence solar-to-chemical energy efficiency in the electrode-type reactors. For crystallized semiconductor materials, size of crystals, surface defects, and possible lattice dislocation act as key factors in determining photoelectrochemical activity. For example, smaller particle or cluster size for photocatalysts leads to shorter travel distance with reduced probability of electron-hole recombination.

Considering the fact the energy levels bend at electrode-electrolyte interface, one can predict that we need to overcome band gap higher than 1.23 eV to activate the photoelectrode for water splitting reaction. Surface properties such as roughness and defects also affect electron-hole recombination processes. Co-catalysts, or promoters are usually introduced to the surface of electrodes, as many metal oxides have low CB for H_2 evolution reaction.

Intrinsically, solar energy conversion involves complicated multistep catalytic reactions. Therefore, key parameters evaluating catalyst performances are also needed to describe photoelectrode in the cells. Turnover numbers (TON) are extensively used in organic synthesis and catalysis to reveal the total number of reactions occurring in the presence of catalytic materials. TON is defined as number of reacted molecules (e.g., water) in the presence of known amounts of active sites. Catalysts can be in either homogeneous or heterogeneous form. To further quantify the rate of catalytic reactions over a period of time, turnover frequency is also introduced, defined as number of reacted/converted molecules over the number of accessible active sites per time.

$$TON = \frac{\text{Number of reacted electrons}}{\text{Number of atoms in a photocatalyst}}$$

$$TON = \frac{\text{Number of reacted electrons}}{\text{Number of atoms at the surface of a photocatalyst}}$$

Sometimes, TON and TOF values can also be expressed as per weight catalyst material. This is very common in industrial catalysis for commodity products. However, it might not be appropriate in this field, as weight of photocatalytic materials cannot reflect the number of active sites on the surface of catalysts. Under other circumstances, photocatalytic activity can also be presented as number of excited photons on the surface of photocatalysts, it however is also dependent on the absorbed photons.

2.4 Nanostructured catalysts: design principles

2.4.1 Criteria for selecting photocatalytic materials

Numerous reports have been published for novel photocatalytic materials with remarkable performances on half reactions. For example, TiO_2, Fe_2O_3, WO_3 and $BiVO_4$ catalysts are among most popular to be found as effective materials for water oxidation or reduction reactions. But very few materials are effective under visible light irradiation. As mentioned earlier, the efficiency of particle-based photocatalytic water splitting systems is relatively lower than electrode-based reactors. In addition, the cost-effectiveness is too low to develop commercially viable technologies as separation of H_2 and O_2 and suppression of back reactions are still facing challenges at current stage. Therefore, developing photoelectrochemical cells to reduce water to H_2 at counter electrode, and oxidation reaction of water at photoanode under minimal applied bias remains a promising option for large scale implementation of photoelectrocatalysis technologies.

External electrical bias can drive electrons from photocatalysts to counter electrode and alleviate recombination of electrons and holes, as well as photocorrosion. To improve STH efficiency is crucial for developing commercially viable technologies. On one hand, how to reduce fast charge recombination reactions remains a grand challenge. On the other hand, charge carriers should be separated as far as possible. Therefore, extensive research efforts have been paid for sophisticated design of photocatalysts and optimization of photoelectron reactors. Selection of photocatalytic materials determines the overall cost-efficiency of photoelectrochemical cells. The following criteria should be considered in developing such materials, to enhance overall efficiency of photo-to-H_2 conversion.

2.4.2 Criteria for designing photocatalytic materials

Heterojunctions. The band gap for a semiconductor determines the adsorption spectrum of solar irradiation. Typically, band gap for good semiconductor catalysts should be in the range of 1.6 eV to 2.5 eV to harvest visible parts of solar irradiation and enhance efficiency of water splitting reactions. For most semiconductors, such requirement cannot be satisfied as their band structures are either below potentials or much larger than 3.0 eV. Band gap engineering or alignment is often conducted to improve the performances of photocatalysts in electrode-based systems. Two or even more types of semiconductor materials are usually combined to form junction structures to alleviate the issues caused by large band gap in a single semiconductor material. In Figure 2-35, three categories of such junction structures are illustrated to reveal the advantages of band gap alignment. In general, a semiconductor catalyst A is combined with a structural/band gap modulator B. They can be either n-type or p-type of semiconductor materials.

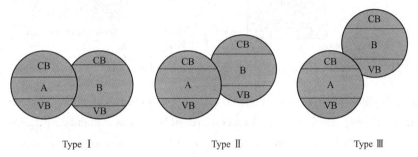

Figure 2-35 Different cases for heterojunctions

In case (a), the band gap of B is larger than A. In other words, VB of B is lower than that of A, while CB is higher than A. As a result, electrons and holes will be generated, transfer and accumulate in A.

For (b) case, there is an overlap in band structures of A and B, where both VB and CB of B is higher than those in A. Therefore, photon induced electrons are generated in B and transfer to A, whereas holes are generated in A and quickly migrate to VB of B. Case (c) is very similar to case(b), except that more pronounced difference in VB and CB positions, gives stronger driving force for charge transfer.

In the past decade, mainstream attention has been paid to the structures of heterojunction structures formed between solid materials. Such combinations include dual semiconductors, semiconductor-metal, semiconductor-carbon types of heterojunctions. The most popular combination is dual-semiconductor type **heterojunctions**. As already mentioned in Figure 2-36 (a), a n-type and a p-type semiconductors form architectures in close contact. The space-charge regions at the interface leads to the formation of electric field driving charge flows, or diffusion of charge carriers. Formation of this type of interfacial junctions enable flows of electrons from CB of a p-type semiconductor to that of n-type semiconductor, whereas holes move from VB of n-type to p-type materials, resulting in more effective charge separation, longer charge carrier lifetime and therefore an enhancement in catalytic efficiency.

Alternatively, for semiconductor-metal type of heterojunctions, a Schottky barrier can be formed at interface when a semiconductor is in close contact with a metal surface. The Fermi level alignment is induced by electron flow from the side with higher Fermi level to the lower one. The metal acts as an electron trap to receive photon excited electrons from semiconductor and reduce

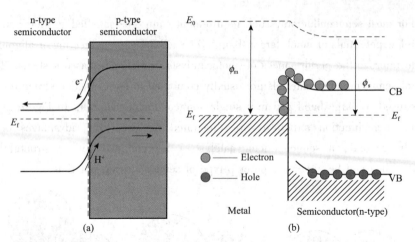

Figure 2-36　Band gap bending and alignment for (a) two semiconductor and (b) semiconductor-metal junctions

recombination, because charge flow cannot occur in an opposite direction. For semiconductor-carbon based interfaces, novel structured carbon materials such as nanotube, graphene are among most favorite owing to their metal-like conductivity and high surface areas, allowing close contact and facile electron mobility from light-absorbing semiconducting materials to carbons.

Band edge positions. To achieve wider absorption spectrum in semiconducting materials, band edges should straddle between the redox potentials of water. As depicted in Figure 2-37, band gap lower than splitting potentials should be visible light absorptive while stable materials with a large band gap absorb lights in ultraviolet region.

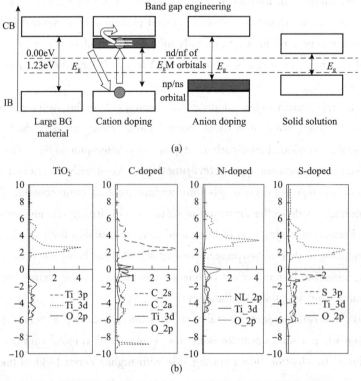

Figure 2-37　Band gap engineering with C, N and S doping

Charge transfer. As mentioned in the above section, band gap engineering is a promising method to improve charge transfer efficiency while reducing electron-hole recombination. Interfacial engineering is carried out to enhance fast charge transfer and prevent loss of photon energy and photocorrosion.

Stability. As photoelectrochemical reactions occur in aqueous medium, hydrophilicity and hydrophobicity properties of photocatalysts is one of the key factors that should be taken into account for long-term efficiency of commercial devices.

Abundant availability. Cost effectiveness and non-toxic nature of photoelectrode materials are important to ensure sustainability.

Band gap engineering. Different from heterojunction strategies to improve charge separation, transfer and prevent recombination processes, doping of cationic or anionic species in existing semiconductor materials can cause alteration of band gap. Doping band gap mediators induces formation of additional levels between VB and CB, or simply modifying CB and VB levels of current semiconductors. For n-type semiconductors, since electrons are charge carriers, facile excitation of electrons from VB can be facilitated by doping additional electron donors, or anions such as N and S. Doping N or S leads to **hybridization** of molecular orbitals with O. Therefore, expansion or widening of VB can be achieved. VB level is closer to Fermi levels of semiconducting materials. Therefore, band gap is optimized by shortening the gap, more absorption of visible lights can be realized to improve **photon-to-electron efficiency**. This process results in negligible changes with CB thus overall efficiency of water splitting can be enhanced within such structure. This methodology is called band gap engineering, where surface-volume property can also be controlled by shape and size control for the nanostructured materials.

Figure 2 – 37 illustrates the mechanism of structural doping in existing semiconductors and modulation of band gap and electronic energy levels resulted from hybridization of molecular levels of TiO_2 as a representative example. Clearly, doping a cationic ion induces distinct band gap structure alignment due to different mechanism involving in orbital hybridization. Hybridization of foreign cations with Ti orbitals expands band levels of CB thus overall energy levels of TiO_2 can be elevated with facilitated electron excitation.

2.4.3 Guidance on catalyst manufacture

One can see that purpose of band gap engineering by doping cationic or anionic species is to shortening band gap for wider spectrum in light absorbance. The mechanistic picture is different from that in heterojunction structures. But on the other hand, one can predict that doping foreign species should also modify the electron-hole separation efficiencies. In light harvesting devices, the following parameters should be considered:

Small particle size. Diffusion depth and region affect electron-hole separation efficiencies. High surface-to-volume ratio determines quantum efficiency. Size-induced band gap alignment plays a key role in surface engineering. Therefore, particle size is one of the key factors governing light absorption processes.

Crystallinity. Nature of crystal phases determine the band gap values for semiconductors. Shape of nanostructure catalysts is considered as another key factor in facilitating band gap alignment as bond length in crystals affect energy levels of CB and VB. High crystallinity leads to high mobility of charge carriers and efficient charge separation.

Cocatalyst. Doping and formation of heterojunction structures contribute to prolonged lifetime of charge separation, preferential migration along a certain direction an efficient chemical reactions with a single product.

2.5 Surface and morphological properties

2.5.1 Size effect

From catalysis point of view, small size means high surface-to-volume ratio and more accessible surface sites for water photon-induced electrons and surface reactions. Low dimensions also favor rapid charge transfer/migration to catalyst surface. Therefore, nanostructured or sub-nanostructured catalytic materials are promising in promoting surface reactions compared with bulk materials. The band bending at electrode-electrolyte interface is significant, particularly for nanosized materials. The charge equilibrium across the interface of electrode causes built-in potential to separate charge carriers. It is particularly true for particle based photoelectrochemical systems. If the particle size is beyond a critical value for space charge region, potential bending is significant (Figures 2 – 38, 2 – 39). If the size is below the critical value, band bending does not exist to separate charges. Therefore, facilitating charge migration while maintaining sufficient charge separation is important for designing nanosized electrocatalysts.

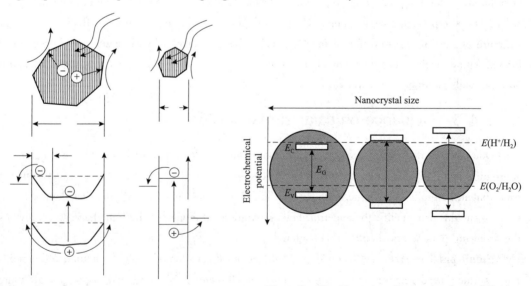

Figure 2 – 38 Nano-size effect on band gap bending

Figure 2 – 39 Nano-size effect on band gap shift

In addition to tuning of energy potentials at electrode-electrolyte interface, the quantum confinement effect is also important. Established in 1980s, quantum size effect is found to be dependent on the shape and size of nanomaterials. A general finding with quantum confinement effect is that, while reducing the size of nanomaterials, CB edge tend to be more reductive and VB will be more oxidative. Such alteration of bandgap could increase light absorption efficiency and minimize surface light reflection thus enhancing light scattering. According to Marcus-Gerischer theory, such modification of bandgap structures intrinsically increases the thermodynamic driving force to enhanced rates for interfacial charge transport and separation.

Studies have also shown a logarithmic dependence of the photocatalytic proton reduction rate on the bandgap on CdSe quantum dots. Clearly, these effects significantly affect performances of photoelectrochemical water splitting efficiency.

The unique properties brought by nanotechnology also include for shortened carrier collection pathways. As will be mentioned in the next section, in the absence of external field, charge carriers move by diffusion and the distance is defined by mean-free diffusion length L. To improve mobility of minor charge carriers at semiconductor-electrolyte interface, surface roughness can be enhanced. This approach is very useful for various metal oxide such as MnO_2 and Fe_2O_3.

Beer-Lambert Law shows that light absorption ability of certain materials is determined by wavelength-dependent absorption coefficient and light penetration depth. To ensure more than 90% absorption of incident light, film thickness is often 2.3-fold larger than absorption coefficient. More importantly, surface structuring on nanoscale can also increase the absorption efficiency (Figure 2 – 40). The trapped light due to high surface roughness would not be lost through direct reflection from a flat surface. Instead, light scattering can be maximized on rough surface.

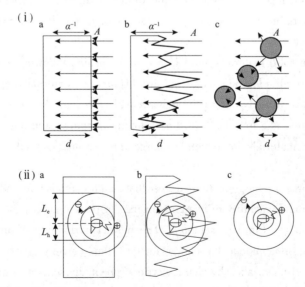

Figure 2 – 40   Electron transfer at nanostructured and flat surface

2.5.2 Diffusion length

While transport of charge carriers can be achieved by applying **external voltage** with close circuit and creating electrical field induced current. Transport mechanism of surface charge carriers across electrode-electrolyte interface plays an important role in surface reactions and quantum efficiency for water splitting reactions. Obviously, charge carriers move from regions where carrier density is high to regions where density is low. The motion of electron/hole carriers is caused by thermal energy and random motion of carriers, rather than electrical field imposed by external potentials. Motion of carriers are not limited to the lattice of electrode materials. We often refer the sitting still or film, or layer as diffusion distance. The charge carriers move towards certain direction with an apparent velocity across the film. When excess of charge carriers are accumulated in a semiconductor, charge carriers diffuses through a characteristic distance. The diffusion coefficient for n and p-types of semiconductors is expressed as:

$$D_{n/p} = D_o \exp\left(-\frac{Q}{RT}\right)$$

For intrinsic semiconductors, diffusion length for electrons is often larger than holes because of larger diffusion constant. Typically, the diffusion length is in the range of micrometer. Both diffusion constant and length are temperature dependent. In particular, the magnitudes of estimated diffusion length increases with decreasing temperature, but decreases with increasing density of dislocations in the absorption layer. Diffusion distance is also strongly dependent on the nature of crystallinity and epitaxy layer. Diffusion length is important, because it strongly affect activation energy on catalyst surface. For example, for silicon based solar cells, the lifetime for charge carriers is approximately 1 msec and the diffusion distance is about 0.1 mm. The two parameters clearly determine the intrinsic performances of electrode materials for water splitting reactions.

Charge carriers in solid materials move in a random fashion following any particular direction any time, thus the overall electric charge is zero. Under an externally applied electric field, electrons and holes move quickly across through a metal or a semiconductor material. The term, electrical mobility, which is a function of material doping, impurity and temperature, describe the properties of electrode materials for effective charge transfer and capability for reducing charge recombination.

Mechanistic pictures of charge motion suggest that, both composition and defects in the solids might affect the measurable mobility of charge carriers. In perfect solid materials, electrons move in the absence of resistance (under vacuum). The perfect periodicity of an ideal solid presents zero resistance and does not impede the charge motion, thus carriers move as a superconductor. However, in actual solid materials, the charge carriers repeatedly scatter off due to crystal defects, surface and polar nature of semiconductors, positioning of substituting heteroatoms in lattices, impurities, *etc*. The presence of scatters changes flow direction of carriers and upset the periodicity of the potential to interrupt the course of charge motion.

If such imperfect structural periodicity occurs at interfaces, e. g., heterojunctions, metal-metal interfaces, semiconductor-metal interfaces, the resistance and scattering happen and due to the trapping of charge carriers. The defects at interface, or roughness on the surface of solid in close contact with electrolyte, increases the possibility that crystal phases become more charged thus alters the course of charge diffusion through the interfacial layers. Different from bulky phase upset in real solids, scattered carriers at interface only moves in two dimensions, the acceleration of which in an external electric field across the inversion layers is disrupted leading to dramatic changes in mobility of charge carriers. **Drift velocity** is introduced to describe the significance of scattering processes (Figure 2-41). Drift velocity is a function of electric field, mass of carriers and scattered time.

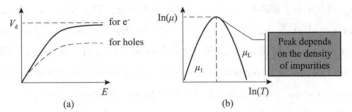

Figure 2-41 (a) Variation of drift velocity with applied potential, (b) Temperature dependency of mobility

Mobility of charges is the sum of electron and hole flow. The values are determined by their corresponding concentrations (number density), elementary charges and electric field. The flow of carriers contributes to electrical conductivity of semiconductor materials. Electrical conductivity (σ) = mobility × carrier concentration. Temperature dependency of mobility is reflected by the randomness of scattering of carriers within the solid materials. Increasing temperature enhances phonon concentration but causes enhanced scattering phenomenon, disrupting carrier mobility. It is believed that mobility of non-polar semiconductors shows a temperature dependency of $T^{-1.5}$, while optical phonon scattering and charged defects are found to be proportional to $T^{-0.5}$ and $T^{1.5}$, respectively. On the other hand, ion impurities may affect mobility in an opposite way. Because increasing temperature would alleviate scattering due to increasing average thermal velocity. This is the reason that mobility shown displays maximum values with regard to temperature.

Identifying the contribution from mobility and diffusion is important to describe the conductivity of semiconducting materials. The transport of current, as already mentioned, is driven by external voltage and concentration gradient of charge carriers. Mobility and diffusion are influenced by electric field and diffusion capability, respectively. Einstein relation correlates the two current transport phenomena using temperature and charge.

$$\frac{D_n}{\mu_n} = \frac{kT}{q}\frac{D_p}{\mu_p} = \frac{kT}{q} = 25\text{mV at room temperature}$$

Impact of doping and impurities on mobility will be discussed in next section in details.

2.5.3 Doping

Substitute O or metal cations in metal oxide materials by metal or nonmetal species will cause a change in the chemical state in existing framework. Doping or substitution is often accompanied with change in the nature of chemical bonds, either compressed/elongated or simply replace original species in lattice structures. Therefore, attractive or repulsive force might be enhanced due to lattice strain. For example, substitutional doping of cations or anions in TiO_2 crystals, leads to changes in local electronic density and reconfiguration of molecular orbitals owing to hybridization of newly added species with existing O or cations network. Sulfurization of CoO, NiO or Fe_2O_3 oxide catalysts alters the nature of metal—O with metal-S bonds resulting in band gap changes in these semiconductors.

Typical substituents include metals (Fe, Cr, V, Mo, Re, Mn, Co, Bi, *etc.*) and nonmetals (N, S, B, C, F). They are proved to be effective dopants for TiO_2, Fe_2O_3, $BiVO_4$ and WO_3 catalyst materials, which are widely used in water splitting. The purpose of doping is to modify band structures for metal oxide materials. It is expected that, a lower shift of CB or a higher shift of VB or formation of impurity band should be achieved in band structures. This part has been discussed in section 2.4 as an important criterion to select or design effective electrode materials. But one should pay attention that, dramatic band gap modification such as gap shortening is not always favorable, depending on the functionality of semiconductors for water splitting reactions. For example, lower level of CB should still be higher than that of H_2/H_2O level, while higher level of VB should be lower than O_2/H_2O level, to ensure occurrence of photo-reduction and photo-oxidation reactions. Generation and transfer of electrons and holes should migrate and react within their lifetime.

In the field of electronic industry, doping is often conducted as regular operation, because introduction of conductive species in semiconductor materials by substituting foreign valent elements for host element. But in photocatalysis and water splitting reactions, the role of doping on increasing total number of carriers by introducing more conductive materials remains to be illustrated for fundamental understanding. In this section, four types of doping such as metal ion doping, nonmetal ion doping, co-doping and self-doping will be discussed in details.

(1) Metal ion doping

Substituting Ti^{4+} with metallic cations is generally found to alleviate the band gap by red shifting the CB level. Both theoretical and experimental studies have suggested that such doping usually alters the band structures by lowering the CB or insertion of a new band closer to CB. The size and charge of dopant affect the resultant lattice structures of TiO_2. For example, many cations are found to be effective to tune the absorption spectrum of TiO_2 catalysts for visible lights. Cr, V, Mn and Fe are among most efficient cationic elements to shift absorption range towards more visible light area.

It has been proposed that the shift towards visible range of solar irradiation originates from the fact that, hybridization between *d* orbital of Ti elements with that of substitutional metal elements, decreasing the band gap of TiO_2 enable wide absorption of lights after implanted with metal cations. For example, substitution of Ti^{4+} with Cr^{3+} or Fe^{3+} cations into original TiO_2 framework form Cr—O or Fe—O bond in the implanted layer, thus decreasing the band gap energy thus enhancing optical responsiveness towards 450~800 nm. Shallow substitution affects band gap, while Fe^{3+} doped TiO_2 prepared by sol-gel methods shows a deep substitution from surface to sublayers. Therefore, the presence of Fe^{3+} also assists charge separation and migration of photo generated electrons and holes by containing the mobility at interfacial level. Although applications varying, from water splitting to degradation of organic compounds *via* photo oxidation, doping metal ions into TiO_2 usually create extra band level near below CB, thus water reduction can be achieved, or formation of anionic O_2^- species causing degrading reaction of pollutants. The synergism of doping can also be realized through tandem catalysis. For example, TiO_2 doped with Rh^{3+} ion can facilitate excitation of photo electrons by Rh^{3+}-CB and VB-Rh^{4+} to complete redox circle (Figure 2-42).

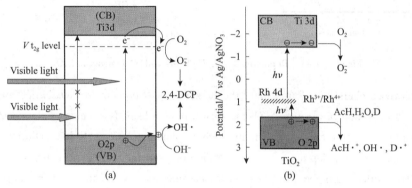

Figure 2-42 Doping of Rh^{3+} in TiO_2 energy band

(2) Nonmetal ion doping

The advantages of metal ion doping including tunable band gap shifting due to altered lattice parameters induced by different sizes and charges imposed by foreign cations. But metal ions are charged impurities, which indeed show limited capability in restraining electron-hole recombination. More importantly, durability is another factor which needs to be taken into serious account. As immersion in aqueous phase may cause structural decomposition due to strong solvation effect imposed by the polar solvent.

Numerous attempts have been made in literature in doping nonmetal ions in metal oxide semiconductors. Since band structures in TiO_2 materials are reflected by d band of Ti and p_π band of O, it is possible that nonmetal ions interact with Ti or O and hybridization processes occur to reconfigure the band structure of TiO_2. Different from the case of cations, doping of nonmetal ions can either induce downward shift of CB or uplifting of VB in metal oxide materials. According to Figure 2-43, B, C, P, S, N and F elements have distinct energy levels across the band structure of TiO_2 materials. The ion radius of those elements is usually similar to O, therefore, **lattice distortion** might not be the determining factor that drives the band gap changes. In the past few years, N doped TiO_2 materials have drawn wide attention. Two possible modes of substitutional states, N—Ti—O and Ti—O—N can be achieved by varying preparation conditions. Interstitial insertion (NO interaction through $\pi*$) could induces formation of extra band levels of 0.14~0.73 eV above VB. The electrons excited to CB could reduce O_2 molecules to generate superoxide anions

(O^{2-}) and facilitate formation of super active oxidative species such as H_2O_2 and hydroxyl radicals.

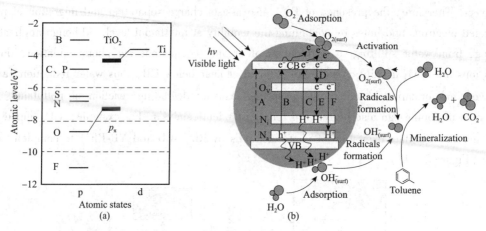

Figure 2-43 Doping of B, C, P, S, N and F elements in TiO_2

Previous researchers have found that, the presence of N content in TiO_2 lattice may affect the generation and migration of electrons. Owing to the dual role of N species, balance of N content and O vacancy is important to maintain higher photocatalytic activity. Diffuse reflectance spectroscopy reveals that, doping N facilitate excitation step for immediate generation of electrons and holes. The surface trapped electrons decreases significantly because, deep trapping induced by O vacancy store charge carriers thus total number of electrons migrating to surface decreases dramatically. Therefore, nonmetal ion doping can influence the kinetics of electron transfer by tunable band gap values and electron transport to one site to another.

(3) Co-doping

Co-doping additional cations or anions exhibit even higher promotional effect on photoelectrochemical performances. The synergies between different doping elements could facilitate charge generation and restrict electron-hole recombination processes. Examples of co-doping include introduction of N—F, N—B, N—Ta and Fe—Ho pairs in TiO_2 catalysts.

Take N—Fe co-doping as a representative example, co-existence of both N and Fe anions alters band gap pictures by forming middle band states. Another example is N—B doped TiO_2. UV vis characterization suggests that absorption of solar irradiation spectrum to 700 nm, varying band gap by 1.94 eV accordingly. A strategy of pre-doping interstitial B and followed by structural doping of N effectively weaken Ti—O bonds thus chemical states of TiO_2 can be modified as needed. N—Ta is another good co-doping combination. As Surface characterization shows that hybridization of $2p$ N and $5d$ Ta could add extra VB edge, which facilitate visible absorption. Coexistence of both Ta and N narrows the band gap by promoting electron transfer from a positively charged Ta to N, leading to formation of Ta—N bond in TiO_2 structures. Therefore, two plausible mechanisms dominate the processes of co-doping processes. Formation of extra band levels which either lower CB or enhance VB promote tandem electron excitation. Hybridization states of co-dopants might tune the band levels for adjustable energy levels and narrowing band gap of metal oxide semiconductors.

Additional effects caused by co-doping have also been confirmed in cationic co-dopants. The cooperative effects due to coexistence of two cations may functionalize in extending visible absorption by narrowing band gap and restraining cluster **agglomeration** to maintain at nanoscale. In Fe—Ho co-doping case, experimental studies have demonstrated that Fe^{3+} cations extend visible light absorption spectrum, while the role of Ho^{3+} seems to be containing particle size thus retarding electron-hole recombination.

The impact on co-doping might be progressive introduction of metal or nonmetal ions. $SrTiO_3$ is known as good photoelectrochemical catalyst material. Introduction of Sr^{2+} to TiO_2 actually weakens existing Ti—O bonds and promote formation of a new perovskite phase. Further doping of other elements such as Ir, Ru, Rh and N show that band gap correction is still the dominant mechanism for improved visible absorption.

(4) Self-doping

The fundamentals for metal, nonmetal ion dopants are based on addition of impurities into metal oxide structures. Differences in ion radius, charges from cations in original oxide materials are the intrinsic driving force for lattice distortion and altered band gap levels. One could foresee possible consequences of impurity doping, including structure decomposition or degradation under extreme conditions. Self-doping involves the coexistence of aliovalent species in original materials. One of the most commonly known examples is the presence of Ti^{3+} in TiO_2 catalysts. Mixing Ti^{3+} and Ti^{4+} in TiO_2 materials has been proven to be very effective in tuning band gap and extend visible absorption spectrum. This is because when is Ti^{4+} converted into Ti^{3+}, the balance in local charges will be modified to compensate an extra O vacancy in TiO_2. It is estimated that incorporation of Ti^{3+} species and formation of localized O vacancy enable 0.75 ~ 1.18 eV below CB of TiO_2, which might be lower than redox potential for H_2 evolution reaction. Therefore, in combination of bulk TiO_2 phase, heterojunction structures can formulate to improve electron excitation and mobility. Experimental results confirm this hypothesis by showing enhanced photocatalytic activity of reduced TiO_2 catalysts for oxidation and H_2 evolution reactions. In addition, presence of Ti^{3+} and O vacancy cause changes of visible light absorption of TiO_2 materials and change of color from white to gray, yellow and black. The defects caused by Ti^{3+} are considered as electron donors and position of energy band level for Ti^{3+} doped TiO_2 is located shallow below Ti^{4+} band. Thus electrons in VB can be easily excited to the extra band level for tandem electron flow for reduction of water for H_2 formation. Electronic transition from occupied Ti^{3+} localized states to CB is key for enhanced visible light absorption.

2.5.4 Surface modification

Surface modification is another type of important process to modify the surface physical and chemical properties by bringing additional characteristics from the original surface of the material. Surface modification agents could be nanoclusters, cations/anions, organic molecules and nonmetallic compounds. In general, heterojunctions could be formed in binary or multi-semiconductor systems. It is considered as another type of surface modification. In this section, we will discuss the role of surface modification species on band gap and catalytic circle for water splitting reaction.

(1) Metal clusters

Band gap. Metal salts and metallic clusters can be effective surface modifiers for semiconductors. Studies have shown that noble metal complexes could afford tunable cation-anion bond for binding hydroxyl species in aqueous medium. Electron transfer occurs between VB-CB and sensitizers. Instead of forming additional band levels within original photocatalysts (Figure 2 – 44), metal ions and clusters could form additional surface for tandem excitation of electrons from their VB to CB, where further photon induced excitation facilitate electron transfer across the interface to CB of original catalyst. But similar to doping, there should be some overlap of VB and CB of metal clusters with original phase of photocatalysts. Holes can migrate easily across the interface and electron-hole recombination is effectively restrained due to fast charge separation at interfaces. Therefore, the presence of surface metal ions or metallic clusters creates additional band levels at interface, rather than intrinsically modify the band gap of original semiconductor materials.

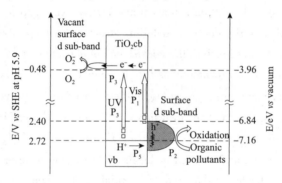

Figure 2 – 44 Metal clusters create sub-band for TiO_2

Plasmonic response. Noble metals existing either in the form of complexes or metallic clusters deposit on the surface of metal oxide catalyst materials have been demonstrated to improve photocatalytic activity in the history. Enhanced catalytic activity is often believed to originate from the prolonged lifetime of charge carriers, as noble metal components are known to be electron sinks, thus accelerating charge transfer from semiconductors to reactant species.

Recently, another important property of noble metal species such as Pt, Pd, Au, Ir, Ru and Ag has brought extensive attention owing to its enhancement in photocatalytic performances. Noble metals are known to activate semiconductors by inducing plasmonic responses on catalyst surface. For example, TiO_2, CeO_2, Fe_2O_3 and ZnO materials can be activated through narrowing the band gap towards wider absorption of visible lights. Plasmon is not a new concept in the field of material science. It has been extensively used for Raman spectroscopy, medicine and optical data storage applications. The systematic studies on the role of plasmonic responses on electron transfer processes involved in photoelectrochemistry only started a few years ago, but with remarkable progress on enhancing water splitting reaction. The mechanism in improved photoactivity and durability is yet to be clarified in existing literature.

Localized surface plasmon resonance for photocatalysis has been established almost a century ago. Plasomonic features have been very well characterized for Au, Ag and Cu nanoparticles. For example, UV-Vis spectra for TiO_2 supported Au catalysts often exhibit chemical shift due to localized surface plasmonic responses and electron transfer between Au and TiO_2 owing to tunable redox properties of both metal and oxide. However, the synergy between plasmonic response and photo-induced electron excitation is still unclear. Therefore, three main mechanisms are proposed for plasmonic photocatalysis under visible light irradiation, (a) electron transfer, (b) energy transfer and (c) plasmonic heating.

The first two mechanisms are still primarily focused on photo induced electrochemical reactions. The last one involves thermal catalysis activated by plasmonic heating localized at surface sites. Again, we take Au/TiO_2 as the most representative example for plasmonic responses in photocatalytic degradation of pollutants (Figure 2 – 45). As the first step, incident photons are absorbed by Au nanoclusters on TiO_2 surface through local surface plasmonic activation. The photo induced electrons are then transferred from Au sites to CB of TiO_2. The electrons reduce molecular O_2 adsorbed on TiO_2 and enable formation of electron deficient Au clusters, which oxidize organic compounds to recover original metallic surfaces.

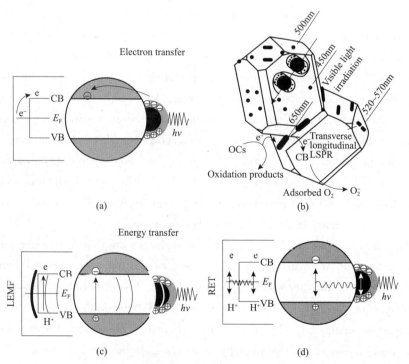

Figure 2 – 45 Origin of surface plasmonic responses

The mechanism of the local surface plasmonic effect has been studied in many previous literatures. Several books and review papers have summarized the comprehensive pictures of possibly each step involved in Au/metal oxide photocatalysts. Only a few points are briefly mentioned in this section.

(a) Absorptive spectroscopy using IR probe can identify the absorption of electrons injected from Au to CB of TiO_2. (b) Different metal oxides have contrasting capabilities for electron transfer and band gaps. (c) Formation of heterojunctions among metal oxide interfaces alter the photo conductivity and affect electrode potentials. (d) Generation of electron-hole pairs and only electrons definitely follow distinct mechanisms. (e) Modification of band structures due to strong adsorption of organic molecules (e.g., sensitizers, pollutants) changes photo absorption spectrum in UV vis range. (f) Detection of radicals in identifying key reactive species under irradiation of UV and visible lights during degradation of pollutants and water splitting.

(2) Nonmetals

Different from surface modification using metal clusters, presence of nonmetallic modifiers has also been found to create additional surface sites and heterojunction structures on the surface of metals or metal oxide. Also different from doping, anions and inorganic compounds are predominantly doped on the surface of catalyst rather than in the sublayers or lattice framework. As mentioned earlier, doping of nonmetal species modified the lattice parameters of metal oxide catalysts leading to the formulation of cationic/anionic vacancy and shortened band gap. Presence of extra metal oxide clusters acts as photo absorption enhancers.

Surface modification by organic/inorganic compounds have been widely reported. Actually, a variety of promoters such as glucose and dyes. Generally speaking, there are two distinct mechanisms which could describe surface modification by nonmetallic species. For modification of semiconductors with dyes, previous studies showed that dye-sensitized solar cells are efficient in transporting electrons under irradiation of visible lights. TiO_2 catalysts coated with a monolayer of charge transfer dye such as ruthenium complexes display enhanced performances. In general, dye-sensitized solar cells consist of the following main components, where electron transfer properties are mediated due to the fact that, (a) adsorption of dye molecule complexes on conducting glass, (b) electrolytes with redox mediators (e.g., I^-/I_3^-). The overall efficiency of solar cells depends on LUMO and HOMO of photosensitizer and Fermi level of TiO_2 electrode, redox potential of mediators. The elementary steps can be described as below:

Photosensitizers adsorbed on the surface of semiconductors are excited from the ground state to excited state under light irradiation. The photo induced electrons are injected into the CB of semiconductor electrode. In other words, dyes are oxidized. Electrons diffuse through the interface of semiconductors and eventually reach counter electrode through the circuit. On the counter electrode, the redox mediator I_3^- obtain electrons to form I^-. The oxidized dyes accept electrons from I^- anions and regenerate to the ground state to complete the catalytic circle.

Reduction of oxidized photosensitizers and I_3^- species could limit the overall efficiency of dye sensitizer solar cells. To improve the performances, novel design of dyes, modification of nanoarchitecture of semiconductors, selection of redox pairs. Since excitation of electrons in dyes occurs instantaneously, diffusion of electrons in the film of semiconductors is rate limiting. Thus increasing crystallinity, better connection between metal clusters and oxides, optimized surface compositions could lead to better light harvesting. Still, hindering of electron and hole recombination by fabricating layered structures is known to enhance H_2 evolution reaction.

Photosentisizers are often used with semiconductors. Such combination effectively improves charge separation under solar irradiation. In these constructs, organic dyes (sentisizers) are excited from ground to excited state, generating electrons injecting into CB of semiconductors to initiate charge separated state at interface of dye-semiconductor. This step is referred as photosensitization. Good dye-semiconductor contact is important to ensure good electronic coupling through strong interfacial binding. The oxidized dye molecules can then be reduced by accepting electrons from a donor in reaction medium (for example, I_3^-/I^-). The secondary electron transfer reduces dyes to ground state and produces a freely diffusing oxidized donor. As a result, charge recombination could be minimized to a large extent. Alternatively, electrons can also flow in undesirable direction and recombine with excited dye molecules. If surface defects exist on the surface of semiconductors, electrons could be trapped for facile charge separation.

As one of the most frequently used moieties to attach organic molecules to TiO_2 surface, is to use carboxylic acids. The carboxylic group bind with metal oxide surfaces via a single bond formed through O atom or a bidentate, or bridging structure. Carboxylic groups can attach efficiently on TiO_2 surface in anhydrous organic solvents. However, the binding is labile in the presence of acidic or protic medium. Other studies also show that, phosphonic acid group is able to form stable attachment in aqueous medium compared with carboxylic acids. The surface modifying agents facilitate electron injection into CB of metal oxide. Such properties are dependent not only on chromophore and anchoring groups, but also the nature of semiconductors such as bandgap and surface roughness.

But it is critical that the stability of TiO_2 particles in liquid medium might complicate the actual performances of photoelectrochemical cells. The dispersion, dye aggregation on metal oxide surface, particle agglomeration, strength of dye/organic molecule-semiconductor and structural heterogeneity in surface binding strong affect the overall performances of cells. To minimize the degree of agglomeration, particle encapsulation methods can be used to create inverse micelles to stabilize dye-metal oxide system. For example, perylene-TiO_2 interface can be formulated in stable fashion. Formation of micelles not only promotes stabilization, but also alters Fermi level of semiconductors with improved electronic transfer properties. The anhydride group is part of the ligand with conjugated chromophoric structure. The modifying agents can also inject electrons into CB of TiO_2 during solar irradiation. The formation of $TiO_2(e^-)$ species is facilitated while charge recombination occurs on a comparatively longer time scale. Therefore, charge separation efficiency is enhanced in this system.

Similarly, photosensitizing $Ru(bpy)_3^{2-}$ complex can act as both light absorber and electron donor with TiO_2 electrode as electron acceptor. IrO_2 functionalizes as a secondary electron donor. In a typical process, light induces electron injection of Ru complexes into CB of TiO_2, leading to the current flow in external circuit and reduce protons to H_2 at Pt electrode. Holes are transferred to IrO_2 species.

For pollutant degradation, electrons at CB of semiconductors can directly reduce the hydrocarbons. Or superoxide species can be generated through charge transfer and eventually oxidize the contaminants.

Electrons accepted by semiconductors can be transported to electrode and externally moved to noble metal electrode to reduce other components. For example H^+ can be reduced to H_2, for HER application. Such reduction process, as inspired by photoelectron chemical cells, can be utilized to produce other fuels and chemicals.

2.5.5 Heterojunctions

Mixed metal oxides, metal oxides/sulfides/phosphates and metal-metal oxide hybrids are representative examples for coupled semiconductor materials for water splitting reaction. The basic principles for hybridizing coupled semiconductors are to restrain the recombination of electron-hole pairs, enhance the diffusion of charge carriers through the thin film on the surface of semiconductor materials. Coupling two distinct electrode materials undoubtedly create interfacial strain and possible lattice reconstruction thus band gap structures are altered, as a variety of examples have been reported for enhanced charge transfer rate, such as CdS/TiO_2, CdS/ZnO, Cd_3P_2/TiO_2.

In the meantime, we have also known that numerous materials already have narrow band gap for solar applications, however, due to the insufficient band gap across the potentials for complete water splitting reactions. But for half reactions, CB levels of semiconductors should be tuned to ensure lower potentials compared with reduction of molecular O_2, allowing efficient consumption of photo induced electrons and degradation of organic molecules with holes for subsequent reactions. For example, some semiconductor materials such as WO_3 have relatively positive potentials than oxidation of water. Therefore, noble metals are usually introduced to create additional band levels which are negative to water oxidation. This is the reason that metal-WO_3 catalysts show good performances for degradation of organic compounds under visible lights. Coupling of two or three different metal oxides has been applied with the aim to restrain electron-hole recombination in semiconductors. In addition to the examples mentioned above, Cu_2O/TiO_2 and Bi_2O_3/WO_3 are also good heterojunction materials for water splitting. Two mechanism have been investigated for coupling wide band gap materials for transfer of one type charge carrier and simultaneous transfer of both carriers.

(1) Excitation of one type of charge carrier

Again, the well-studied TiO_2 materials are often known as electron acceptors due to wide band gap levels. Doping or hybridizing sulfides with TiO_2 nano-belts and tubes exhibit enhanced features for absorbing visible light spectrum. The well-structured heterojunctions between TiO_2 and sulfides also enable prolonged life time of charge carriers before recombination occurs.

Such hetero-interfaces can also form between organic and inorganic surfaces, as $\pi \rightarrow \pi^*$ transition can occur for $g-C_3N_4$ and TiO_2 interfaces. LUMO of is relatively negative than CB of TiO_2 thus photo induced electrons can be injected into CB of TiO_2 easily.

(2) Excitation of both types of component

Different from the case with sulfide/TiO_2, some hybridized semiconducting materials show synergistic dual excitation behaviors. For example, VB levels of TiO_2 are determined by O $2p$ orbitals, while incorporating other semiconductors with more negative VB levels than TiO_2. Therefore, both semiconducting components are exposed for photo excitation.

For perfect interface with negligible defects, coupled n and p types of semiconductors can efficiently separate electrons and holes by forming junctions to ensure charge flow. However, in most cases, mixed composites of n and p types of semiconductors with defected interfaces

2.6 Standards for experimental studies

2.6.1 Catalyst preparation

(1) Preparation of semiconducting materials

Preparation techniques determine the performances of semiconductors in photoelectrochemical water splitting reactions. Both advantages and drawbacks should be kept in mind for fabricating semiconductor catalysts. Atomic layer deposition by chemical vapor methods can provide good control of surface properties and layer thickness thus utilized in many scenarios. The cost, stability of materials as well as scalability of fabrication techniques are important to be considered when selecting a method.

Chemical vapor method often refers to the evaporation and sputtering processes performed in vacuum chambers. The solid substrate is heated to yield a vapor that condenses on solid surface to form electrode materials. Electrochemical deposition involves a conductive substrate placed in an electrolyte solution. When an external electrical potential is applied between the substrate and a counter electrode, redox reactions occur at the surface of the substrate which deposits materials. Varying electrical pulse or frequencies might sometimes be used to alter the redox reactions thus desired surface or crystallinity could be achieved during electrodeposition.

Crystallinity plays the key role in determining the band gap structures of semiconductor catalysts. The Fermi levels and mobility of charge carriers are highly dependent on the fashion of catalyst preparation and resultant structures of the thin film. Obviously the selection of deposition methods affects the amorphous, polycrystalline, or monocrystalline properties of the thin films. Chemical vapor methods are often preferred in obtaining highly crystalline semiconductor materials. But those methods are costly for scalable applications. Therefore, seeking cost-effective methods for deposition processes are critical for large scale implementation of photoelectrochemical cells.

As mentioned in the previous sections, diffusion length is another important factor for semiconductor electrode materials. To optimize the charge carrier generation, collection and separation processes, the thickness of thin films must be on the order of optical penetration depth (α^{-1}, where α is the absorption coefficient). The thickness of the film often corresponds to 63% of absorption of incident light. Extra thickness of thin film cause excess Ohmic resistance.

Typically, 1 ~ 2 μm thick film is generally sufficient for visible lights in direct solar cells ($\alpha \sim 10^6$ cm^{-1}). However, for indirect cells, much thicker film is required due to low absorption coefficient ($\alpha \sim 10^4$ cm^{-1}). However, materials with poor charge mobility need to be thinner than optical penetration depth for optimal performances. Nanostructured thin films with pore channels that expose the conductive substrate to the electrolyte lead to shunting.

Energy loss due to poor contact between semiconductors and substrate leads to Ohmic loss. Ohmic contacts do not rectify the injection of minority carriers into the bulk phase of semiconductor but can have a moderate resistance. Values of work function are important in selecting contacting materials. Au is often a good candidate for large work function Ohmic contact, particular for p-type semiconductors. In this configuration, holes gather at the surface forming an accumulation layer and semiconductors behave like a metal at the junction.

For n-type semiconductors, Al is a small work function candidate. It is worth pointing out that, selection of a contact material can be guided by the reported work, but best done with careful characterization of electronic properties of the semiconductor/substrate interface.

Connecting semiconductor materials to external circuit depends on the setup of testing kit. When a conventional laboratory open beakers or glass cells are used, a wire must be attached to the conductive substrate upon which the semiconductor materials are deposited. Such connection needs soldering the wires on the conductive substrate using Ag paint and depositing In. But glass materials coated with FTO substrates, the nonconductive glass needs portion of FTO substrate to be uncovered by semiconductors.

(2) Electrode connection

Electrode preparation needs to pay attention to the following aspects: A large contact area is needed to ensure Ohmic contact and uniform potential distribution. Insulation of electrical and chemical for all conductive hardware such as wires, contacts and electrode from electrolyte need to be ensured (Figure 2 – 46).

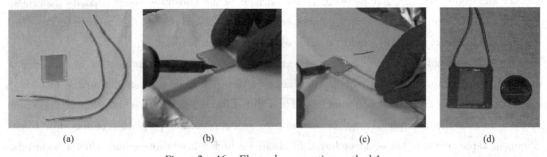

Figure 2 – 46 Electrode connection method 1

One of two wires may be needed to achieve a uniform potential distribution and enhanced charge collection. Both ends of the wire need to be stripped and one end will be soldered onto the outer edge of the uncoated conductive part of the substrate. In is usually used as solder. Epoxy needs to be coated on electrode at room temperature. The electrode can also be connected with a Cu wire by Ag paint (Figure 2 – 47).

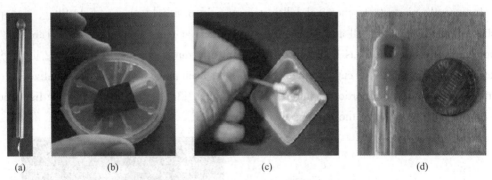

(a)   (b)   (c)   (d)

Figure 2-47 Electrode connection method 2

(3) Electrode surface area

Performances of electrode materials need to be normalized to the planar projected electrode surface area. Therefore, accurate determination of surface area of electrode is very important. Two approaches are often employed to measure surface area. Surface areas can be determined by counting the number of pixels on a digital picture of an electrode. A ruler is needed as reference scale. Some image software can be applied to assist accurate determination. Another method is measuring the geometric surface area within the O ring. The exposed planar geometric area of the semiconductor samples to electrolyte can be determined by measuring the limited area within the O-ring.

2.6.2 Reactor setup

(1) The setup

In the case with three electrodes, double compartment setup with working electrode separated from counter electrode is needed (Figure 2-48). It is also possible that reference electrode be immersed in a compartment connected with salt bridge. The configuration and variation have a great impact on device performances.

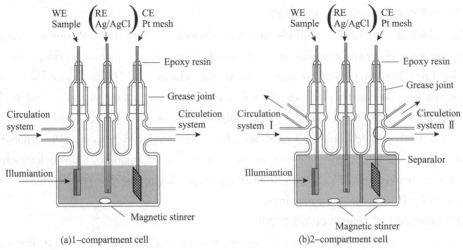

(a) 1-compartment cell   (b) 2-compartment cell

Figure 2-48 Single and dual cell devices

In general, photoelectrochemical cell systems consist of a light source, light filter, a chopper or shutter and the reactor cell and an electrical measurement tool (Figure 2-49). Tungsten and Xenon bulbs are often used as light source for their wide range of energies including ultra violet spectral region. A series of cut-off filters can also be used to achieve a representative spectral distribution. To lower the portion of infrared light, a water filter is placed between the light source and reactor cell to minimize the heating effect.

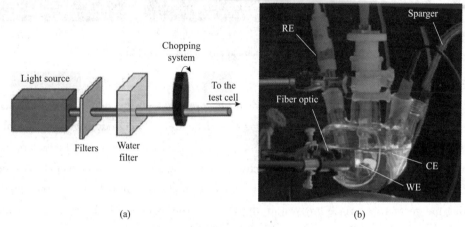

Figure 2-49 A photoelectrochemical cell system

Two reactor setup systems are often used for photoelectrochemical testing. One type includes a cell containing a working electrode and a counter electrode (two electrode system). Another usually have three electrodes, working electrode, counter electrode and a reference electrode.

The reactor cells need to be made with transparent materials to illumination spectrum of interest. If UV irradiation is required, quartz windows must be used in the cell, as regular glass materials will absorb most energy with wavelengths below 360 nm. Furthermore, the distance among electrodes should also be minimized to lower the chemical and electrical resistance of electrolytes during photoelectrochemical tests.

(2) Selection of counter electrode

The selection of counter electrode materials depends on the type of working electrode. Oversized counter electrode materials are often used to ensure reactions at working electrode is kinetic limiting. For n-type semiconductors as working electrodes, H^+ is reduced to H_2 during reduction reactions at counter electrode. Pt foils can be used as counter electrode. But for p-type of semiconductors as working electrodes, H_2O molecules are oxidized into O_2 at counter electrode. In this case, RuO_2 and IrO_2 are good candidates for oxygen evolution reaction. Those noble metal materials are obviously not cost effective for large scale implementation of photoelectrochemical cells. Thus numerous research efforts have been put to developing inexpensive materials to replace noble metals for H_2/O_2 evolution reactions.

(3) Selection of reference electrode

Reference electrodes are necessary in photoelectrochemical testing, to measure the potential of the working electrodes. Among various options, saturated calomel electrodes and Ag/AgCl reference electrodes are easy to use in acidic medium. For basic medium, both of the electrodes will undergo damage due to high pH values. Hg/HgO reference electrodes are preferred in alkaline medium.

(4) Measurement reproducibility

Before activity tests, potentials of individual reference electrode need to be measured to ensure stability. Such operation should be conducted in an open circuit against a master reference electrode. To limit the potential drift between each experiment, reference electrodes are usually stored in the solution recommended by the suppliers. In addition, although potentials of electrode are characterized at room temperature, temperature drift induced by heating may occur during experiments.

(5) Selecting electrolyte

Appropriate electrolytes need to be selected to match the performances of electrode. It is critical to maintain the stability of electrode to resist corrosion when immersed in the solution of electrolyte. Stability is the main concern in selecting appropriate electrolytes. Neutral pH is good to start with for unknown materials although acidic or basic medium often show better performances for H_2/O_2 evolution reactions. The electrolyte should not show strong absorption in the spectral window of semiconductor catalysts. Concentration of electrolyte should be sufficiently ionically conductive to ensure minimal current resistance.

For two electrode systems, inert gas should be injected and electrolyte be sparged before use. For three electrode systems, corresponding gas should be used to sparge p and n-types of electrode before testing. In some cases, surfactants can also be added to minimize bubble sticking on the surface of electrode to ensure well-dispersive bubbling during reactions.

2.6.3 Light spectroscopy

To study the catalytic properties of semiconductors for H_2 and O_2 evolution reactions, spectral distribution and intensity of lights should be modulated to ensure consistency. Since spectrum and intensity of sunlight are regionally and temporally variable, it is important to understand and control the spectrum and intensity of lighting sources, investigating the surface and structural features of catalytic materials.

(1) UV visible spectroscopy

For semiconductors, UV visible spectroscopy studies provide a very convenient method to estimate the optical band gap levels, as the electronic transitions between VB and CB is reflected by wavelength and energies of photons. Optical band gap is not necessarily equal to the band gap values, as defined as difference between the minimum CB level and the maximum VB level. But it is often approximated to represent the band gap values.

To establish the definition for photon-to-current efficiency, a few conceptual terms should be introduced to illustrate the fundamentals on UV visible spectroscopy. According to Beer-Lambert Law, The fraction of light after interacting with samples, known as transmittance or reflectance, I, against the incident intensity I_0, is dependent on the path length of light through the sample (l), absorption cross section, and the difference between initial state (N_i) and final state (N_f) of electronic energy levels.

$$\frac{I}{I_o} = e^{-\sigma(N_1 - N_2)l}$$

The above equation can also be expressed as Beer's Law:

$$A = \varepsilon c l = -\log_{10}\left(\frac{I}{I_o}\right)$$

A is the absorbance. is the molar absorptivity coefficient of the material. c is the concentration of the absorbing species, and l is the path length of light through the sample.

(2) Methods for performing UV visible spectroscopy measurement

Several configurations for UV visible measurement are available, such as transmission, diffuse reflectance and absorption. Each of these techniques need to follow a standard protocol in measurement.

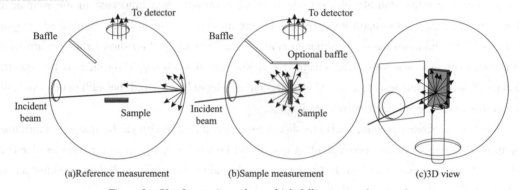

Figure 2-50 Integrating sphere which fully encases the sample

Researchers should decide which of the above mode will be conducted for measurement. In general, transmission mode is used for samples which show some extent of transparency. These materials are thin films supported on transparent substrates. But for thick samples with poor transparency, for example, supported metal substrates, transmission mode is not suitable as the spectrometer would not receive signals. Under this circumstance, opaque samples should utilize a diffuse reflectance or absorption configuration. Diffuse reflectance mode is also useful for particle or powder samples, Integrating sphere is often employed to provide greater signal compared with the measurement under absorption mode.

In transmission mode, the samples were placed on the holder, referred as working samples, in the pathway of a collimated beam of light. Samples should have some degree of transparency. The light impinges on the working sample, and is partially absorbed at certain characteristic wavelengths, according to the electronic configuration and transition in the sample. A spectrometer collects the transmitted light and compares the output with a baseline measurement which is referenced as 100% transmission. The reference measurement should be obtained in the presence of support, such as cuvette holder or a glass plate. The transmission reference measurements can be accomplished using a one-beam or two-beam setup.

For one-beam setup, one should conduct baseline scan by placing reference sample in the path of the beam first. Then the user can replace the reference sample with the working sample for measurement (Figure 2-51). The main issue plaguing with such setup is the potential drift of beam during the course of measurement and sample switching, particularly during the warm-up process. Therefore, the measurement of reference sample is often carried out immediately prior to working samples to minimize the possibility of fluctuation which might occur. Two-beam instruments are available to ensure a dynamic baseline scan.

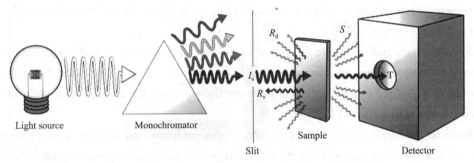

Figure 2-51 Transmission mode of measurement in UV-Visible detector

In a typical configuration for a diffuse reflectance measurement, the setup includes an integrating sphere which collects the reflected photons from the sample. In general, the sphere is placed with an inlet port connected with the lighting source, and an outlet port which is linked with the detector.

The inside wall of the integrating sphere is coated with highly reflective materials such as PTFE or Ba_2SO_4. Such materials could potentially reflect lights with a wide range of wavelength. Two types of reflection can occur during measurement, specular and diffuse. Specular reflection occurs when the beam light strikes a sample and reflects at an angle which is equal to the angle of input. Specularly reflected lights has not undergone significant interaction with the working sample, although partially absorbed thermally. Therefore, it has limited information on the electronic states of working sample.

In contrast, in diffuse reflectance, the incident light penetrates the sample, partially absorbed and only a fraction of excited photons is reoriented (reflected) through various angles. During diffuse measurement, the specular reflectance will increase noise and decrease the accuracy of diffuse measurement. For particulate samples, dilution of samples in non-absorbing matrix is necessary. Typical non-absorbing materials are KBr, KCl and Ba_2SO_4. Different from absorption measurement, the collection of reference substrates is very important for accurate measurement of working samples. For samples deposited on reflective substrates such as metallic and metallic molybdenum, the metal itself serves as a reference to account for all absorption. For samples loaded on transparent substrates, use of a white diffuse reflectance standard needs to be followed strictly.

The wavelength window for various UV-Visible measurement instrument ranges from UV to near IR spectrum. If the light source is monochromated, the sample is under irradiation of a single wavelength during the course of measurement. The transmitted or reflected light is detected with signal enhancing device. Alternatively, if the sample is exposed to broadband illumination, the collected signals can be monochromated and sent to detector. In this case, the detector can capture light data at all wavelength simultaneously. Clearly, based on the description for absorptive and reflective measurement, the following setup needs to be assembled in a laboratory: light source, monochromator with filters, integrating sphere, suitable detector, and optical fibers.

The selection of lighting sources is determined by the purpose of measurement. For typical UV radiation measurement, 160 ~ 400 nm of deuterium lamps are preferred for measurement. For an extended range of wavelength from 200 ~ 2500 nm, xenon and mercury lights are wanted in measurement. Light filters is crucial in eliminating stray light which increases rapidly when wavelength is lower than 400 nm.

The size of integrating sphere is critical for precise measurement. Smaller spheres result in brighter light output. However, incorporating input, output and sampling ports into one small sphere might distort the original spherical geometry thus interfere the hemispherical reflectance.

(3) Band gap analysis from spectra

Based on the theory, for a perfect surface of semiconductor, no absorption for photons with energies below the band gap will occur, while sharp increase in absorption of photons will take place for photons with energies above the band gap. The typical wavelength and energy conversion is shown below:

$$h\nu(eV) = \frac{1239.8(eV \times nm)}{\lambda(nm)}$$

The above equation can describe the minimum energy required for electron excitation to a specific point. However, real spectra exhibit a strong non-linear increase in absorption which reflects the local density of states for CB and VB. Other factors, such as local defects could also contribute to the measured band gap.

For actual measurement, transmission, diffuse and reflection all contribute to the ligh intensity measured by the instrument. These effects reflect the morphology of each sample, which could complicate the analysis of band gap and electronic configuration due to imperfect surface. One can correct those influence by shifting data points of the lowest absorbance to near-zero baseline. As a result, the reflectance and scattering are wavelength independent.

Band gap analysis is often involving plotting and fitting the absorption data. Absorbance A can be normalized to the path length l of the light through the sample and absorption coefficient α. The following Tauc relation is presented.

$$\alpha h\nu \propto (h\nu - E_g)^{1/n}$$

Where n have the value of 3, 2, 1.5 or 0.5, ascribed to indirect (forbidden), indirect (allowed), direct (forbidden) and direct (allowed) transitions, respectively. The Tauc plots of $(\alpha h\nu)^n$ against $h\nu$ yield the value of the band gap when extrapolated to the baseline. It is important to be aware that, the value of absorption coefficient α might affect the error generated during calculation of band gap using the above equation, although it is possible to assess the band gap value without normalizing A to α. For diffuse reflectance measurement, Kubelka-Munk model should be used to estimate α, considering the influence of scattering.

$$f(R) = \frac{(1-R)^2}{2R} = \frac{\alpha}{s}$$

2.6.4 Flat band potential techniques

(1) Open-circuit potential

Fermi level energy can be estimated by illuminated open circuit potential, if the band gap illumination is intense enough to remove existing band bending at the interface of electrode and solution. In addition, the material does not show fast rates for charge recombination.

As already mentioned in the previous sections, the immense of electrode in the electrolyte solution cause interfacial band bending. But sufficiently strong illumination can remove such bending to large extent. Therefore, measurement of Fermi level energy is possible using open circuit potential.

To determine the open circuit potential in a three-electrode electrochemical cell, connect the voltage to the semiconductor and reference electrode (RE). One way to estimate whether the light intensity is sufficient to flatten the band bending is to plot open circuit potential with illumination intensity. For electrodes which do not show significant band bending, a sufficiently strong illumination would saturate the open circuit potential, after which more intense illumination would not affect the potential values.

To check the conductivity type of semiconductor materials, one should pay attention to the direction of potential shifting under illumination. If the open circuit potential is shifted towards positive values, the material is p-type. If the potential shifts to a more negative value, the semiconductor is n-type. However, if the materials do not show any potential shift during illumination, there might be some issues with electrode contact.

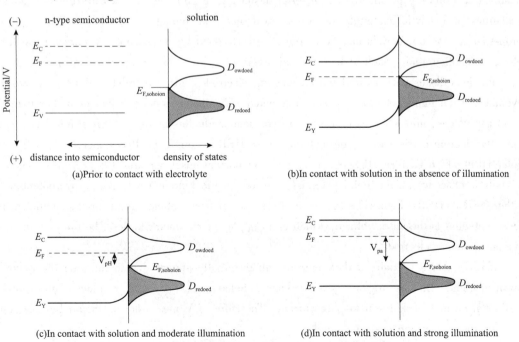

Figure 2-52 Band diagrams of n-type semiconductor

The values of pH significantly affect the open circuit potentials. Fermi energy of numerous semiconductor materials show a Nernstian relationship as a function of pH. The most common way to analyze data is to plot the measured Fermi energy along with OER and HER potentials under different pH values. For example, for a p-type semiconductor material, Fermi levels are more negative of the OER potential. As a result, there is a major energetic barrier for hole injection at the counter electrode (n-type). Therefore, an additional bias is required to overcome the energetic barrier to enable water oxidation reaction.

(2) Mott-Schottky

Mott-Schottky relationship (M-S) can also describe the capacitance of the space charge carriers of electrode as a function of applied potential and Fermi energy.

$$\frac{1}{C_{SC}^2} = \frac{2}{\varepsilon_r \varepsilon_0 A^2 e N_{Dopant}} \left(E - E_{fb} - \frac{kT}{e} \right)$$

Where ε_r is the relative permittivity of the electrode, ε_0 is the permittivity in vacuum, A is the surface area, e is the charge of an electron N_{Dopant} is the free charge density, k is t Bolzmann constant, T is the temperature and E is the applied external potential.

M-S relation is useful to estimate the Fermi energy level and free charge density if the measurement is successful. Fermi energy, band gap and charge density are critical factors to describe the band structure. We need to plot $1/C^2$ against polarization to obtain information. For p-type materials, the M-S plot shows a negative slope while for n-type it is positive.

(3) *j*-V and photocurrent onset

Evaluation of semiconductor materials for water splitting is important under simulated solar irradiation. From the perspective of material design, it is important to measure the fundamental properties of catalytic materials for water oxidation and reduction reactions. The important properties for those materials can be quantitatively assessed by measuring *j*-V curves in a three-electrode configuration. Such technique is also very useful to evaluate the flat band potential.

Previous discussions showed that immersing electrodes into electrolyte solution induces the evolution of flat band potential. When illuminated with photons energy equal or higher than the band gap values, minority hole carriers in n-type semiconductors facilitate OER at interface, while minority electron carriers in p-type materials drive HER at interface. It is important to define the potential at which OER or HER occurs as the photocurrent onset potential (E_{onset}). This potential actually is offset to the flat band potential, caused by the kinetic (activation) overpotential for either OER or HER (Figure 2-53). The difference between photocurrent onset potential to the redox potential is the onset voltage, which is an important parameter to describe the performances of semiconductor electrode.

If reverse bias is high, a dark current will eventually appear due to lowering the Schottky barrier height. In this region, semiconductors behave like a metal conductor and provide additional current in addition to photocurrent. Therefore, *j*-V measurement should be conducted between dark onset and onset potentials.

To further illustrate the metrics of electrode which is used to estimate the performances of solar cells, cyclic voltammograms (CV) curves are used. On a reversible hydrogen electrode scale

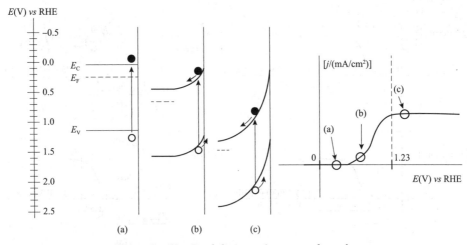

Figure 2-53 Band diagram of a n-type electrode

(RHE), zero voltage represents the potential for HER while 1.23 V represents that for OER. Current density at the potential for HER and OER should be marked for measurement. The voltage difference between actual potentials for OER/HER and theoretical values should also be noted as metrics to describe how photocurrent responses with potentials.

There are eight possible scenarios for CV curves (Figure 2-54). (c) describes a case when photocurrent of a photoanode is overlayed with the dark current response of a cathode catalyst. Since the complete device is operated driving the same current through anode and cathode to maintain overall charge neutrality, the intersection between the two curves represents the theoretical j value. (d) shows a case with photocathode and anode. (e, f) presents a case with both photocathode and photoanode, where the semiconductors cannot generate sufficient onset voltage to drive simultaneous water splitting as j_{SC} is zero in this case. (g, h) show the cases with tandem cell where a photocathode and photoanode are stacked in series for water splitting. The intersection of the two photoresponse curves represent the theoretical values of j_{SC} for the device.

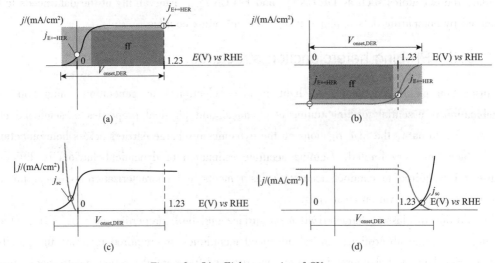

Figure 2-54 Eight scenarios of CV curves

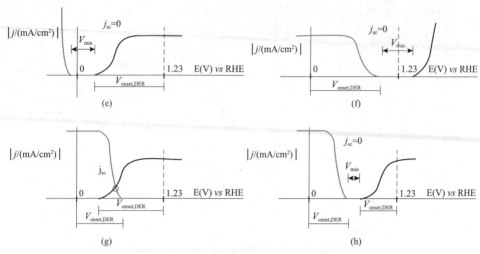

Figure 2-54 Eight scenarios of CV curves (continued)

As onset potential is defined to offset the flat band potential, one should know that, actually other interfacial effects may delay the onset to a point beyond the transition from accumulation to depletion. Such errors could be the in order of mV to a few voltages. Activation potential could also contributes to the shift of onset potential. This overpotential shifts the photocurrent onset in the cathodic direction for p-type samples and in the anodic direction for n-type samples. This is the reason that, catalyst materials are often coated on the surface of electrodes to minimize the overpotentials. However, such modification may also affect the intrinsic junction structures thus shifting flat band potential adversely.

The results of analysis involve a plot of photocurrent density as a function of potential measured against reference electrode. The dark and illuminated behavior of an electrode can be characterized in a single sweep by chopping the light at a certain frequency. Due to the complication of activation overpotential and surface defects on the measured values of flat band potential, redox couples such as $Fe(CN)_6^{4-}$ and $Fe(CN)_6^{3-}$. The validity of the data needs to be considered by comparing data obtained using different redox couples.

2.6.5 Probing heterojunction structures.

Inspection on the dynamics of light induced electron/hole generation, migration and recombination is essential for fine tuning of chemical and physical properties of semiconductor materials. Up to date, the investigations on the dynamics of charge carriers across heterojunctions and interfaces are very limited. Gaining accurate estimation of dynamic behaviors is difficult, because it is required to conduct accurate measurements and characterization to reveal the H_2 production kinetics on various materials.

In addition to classic characterization on surface and bulk properties such as XRD, TEM, XPS and UV visible absorption, several advanced techniques are required to probe the possible charge carrier dynamic behaviors, including time-resolved transient absorption spectroscopy, time-resolved diffuse reflectance spectroscopy, Kelvin probe force microscopy and surface photovoltage instrument. Chemical impedance spectroscopy and Mott-Schottky analysis are also widely used.

(1) Time resolved spectroscopy

Time transient measurement is useful to assess the life time of exciton. For example, anatase and rutile TiO_2 show different lifetime for excitons. The charge pair separation across anatase and rutile were tracked on submicrosecond timescales. For other heterojunction structures, the difference in exciton lifetime can be estimated to probe the band alignment due to formation of interfaces. Such analysis is often combined with XPS and UV visible absorption spectroscopy to reveal the heterojunction structures.

(2) Kelvin probe force microscopy and surface photovoltage spectroscopy

As another important technique to estimate the charge separation dynamics, Kelvin probe force microscopy were used to probe the build-in electric field at the interface of two crystal phases. The electron transfer in such field can be obtained by conducting surface photovoltage spectroscopy. As shown in Figure 2 – 55 (a), the electric field is visualized and electron transfer mechanism can be revealed.

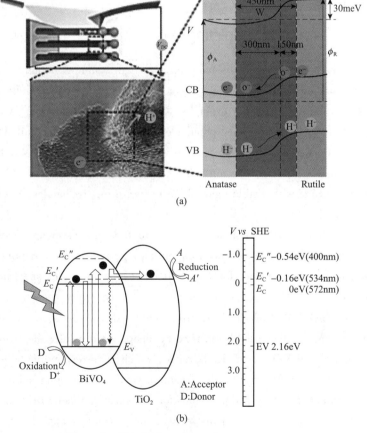

Figure 2 – 55 Surface photovoltage spectroscopy

(3) Chemical impedance spectroscopy and Mott-Schottky analysis

Impedance spectroscopy analysis was used to probe the Schottky heterojunctions. The Mott-Schottky analysis and I-V measurements can provide backup support for the efficient charge pair separation by providing, respectively, the charge density and the current-voltage values. For example, for Pt/TiO_2 with rich oxygen vacancies, the photo induced electrons transfer from bulk to surface of support was facilitated by Pt—O bond bridges. Nyquist plot confirmed that Pt—O bridges can provide higher photovoltage and charge density with lowered resistance of charge transfer across the interface.

2.7 Overall water splitting

Water splitting involves water oxidation and reduction as two half reactions. Water splitting catalysts are classified as one-step and two-step photoexcitation. In one-step processes, light harvesting units are illuminated under solar energy to generate equimolar amounts of electrons and holes, while in a two-step scheme, two equimoles of electrons and holes are generated with one mole for water splitting reactions.

2.7.1 One-step water splitting

In a one-step scheme, CB should be more positive than water reduction potential with more negative VB than water oxidation potential. Semiconductors with wide bandgap structures are suitable for one-step water splitting. For example, ZnS (3.6 eV), TiO_2 (3.2 eV) and ZnO (3.2 eV) and MoS_2 (1.75 eV) are among most popular candidates. However, detailed studies on these materials reveal that, only a small fraction of solar energy can be utilized, as only UV (4%) of solar energy can be used to excite electrons. To better utilize large fraction of solar energy for photocatalytic water splitting with visible light (44%), narrowing the bandgap of original materials is important in this field.

Surface and structural modification are effective to tune the bandgap. Various techniques mentioned in previous sections, such as metal cation doping, non-metallic doping, self-doping as well as size-confinement are necessary for improved harvesting for wider spectrum. Synthesis of solid solution of binary of trinary semiconductor materials is generally acceptable technique. For example, combining ZnO with GaN leads to narrowed bandgap. Because electronic interaction between Zn (3d) and N (2p) causes decrease in bandgap. As a result, the adsorption spectrum is extended from near UV to 450 nm in visible range, which is corresponding to a bandgap of 2.5 eV. Upon loading with Ru or Rh species, the quantum yield can be enhanced as high as 5%.

Figure 2–56 presents bandgap and potentials for various single metal oxide species for overall water splitting. Obviously for many inexpensive materials, bandgap engineering is necessary to narrow the bandgap thus increasing light harvesting efficiency.

(1) Design of binary metal oxides containing cations with partially filled d-orbitals exhibit smaller bandgap due to facile d-to-d transition. Take hematite as an example, it has a bandgap of

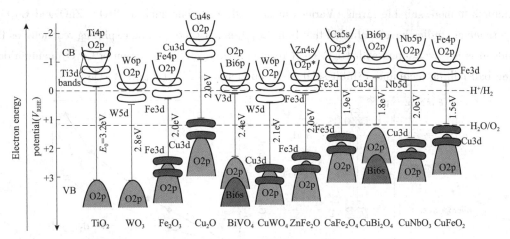

Figure 2-56 Band gap and potentials for various single metal oxides

2.0~2.2 eV with a valence band position suitable for water oxidation. But fast recombination of electrons and holes and poor conductivity cause severe problems with this type of material. Therefore, the main research efforts are focused on enhancing hematite as a photoanode.

Electron-hole recombination is another issue which needs to be solved. Doping noble metals such as Pt and Ru can effectively facilitate charge separation for HER and OER respectively. More importantly, noble metals can quickly remove electrons from the surface of electrode thus minimizing the unwanted reduction of sulfide to sulfur.

Another cost effective material is Cu_2O. It possesses a suitable bandgap structures for photo-induced charge transport and separation properties, which can be used for effective light harvesting for water reduction. With a bandgap of 1.9~2.2 eV, Cu_2O can absorb large fraction of visible light spectrum. This is the reason that it has been widely studied for HER as a p-type semiconductor. However, the electrochemical potential to reduce Cu^+ to Cu^0 is more positive than water reduction, making it quickly deactivated in aqueous electrolyte solution.

Doping ions has been discussed in previous sections. This technique is very important for one-step water splitting processes, as such surface modification is effective for many nanostructured catalysts. Figure 2-57 presents an illustrative case for ion modifying bandgap evolution and energy potentials for water splitting.

(2) In addition to doping, size confinement also leads to tunable bandgap

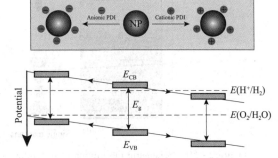

Figure 2-57 Effect of potential determining ions on nanocrystal energetic

structures even for same semiconductor materials in the absence of any other additives. Reducing particle size of semiconductors to nano range leads to strong quantum effects thus changing the potential levels for CB and VB. If bulky semiconductor materials do not have suitable band potentials for water splitting, reducing the size of materials to much smaller size may shift band structures to more suitable levels. Various metal oxide materials such as WO_3, ZnO and Co_3O_4. Experimental studies have confirmed that bulk Co_3O_4 is not good for water splitting reaction, as the bandgap is not overlapped with HER/OER reaction. However, preparation of Co_3O_4 quantum dots using ball milling technique exhibit solar-to-H_2 efficiency of 5% (Figure 2-58).

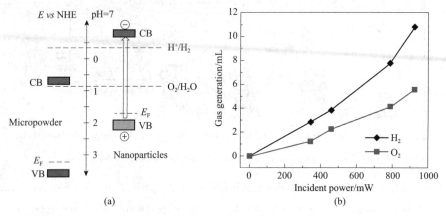

Figure 2-58 Nano effect of Co_3O_4 catalyst

(3) Multiple exciton generation is another promising technique to enhance incident light harvesting. It is a process where multiple electron-hole pairs are produced upon absorption of a single photon on the surface of semiconductor materials. Multiple exciton generation actually proceeds through ionization and coherent superposition of single and multi-exciton states, and multi-exciton formation through a virtual exciton state (Figure 2-59). Such method is a promising route to enhance solar conversion efficiency in single-junction cells. Representative examples for this system include PbSe, PbTe, CdS as well as carbon nanotubes.

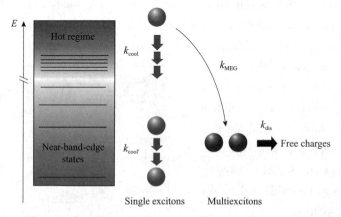

Figure 2-59 Schematic description of relaxation pathways or photoexcited hot single excitons

2.7.2 Two-step water splitting

Two main drawbacks exist for most semiconductor catalytic materials, unsuitable band potential for either water oxidation/reduction and wide bandgap with poor utilization efficiency of visible spectrum of light. Improved efficiency in converting visible range of light ($400 < \lambda < 800$ nm) remains necessary for large scale production of H_2 based on water splitting technologies. Therefore, widening the absorption band is the main focus in current studies. Combining two wide-band semiconductor catalysts in one redox pair of electrolyte can be effective in addressing this challenge. The pair of electrolyte show have the following properties, one with VB below the oxidation potential of water, and another with CB just above the reduction potential. This scheme is defined as Z scheme or tandem photocatalysis, or two-step scheme. Such scheme is actually inspired by natural photosynthesis processes in plants, where photocatalysts can harvest ~700 nm light for oxidation of water to O_2 with extremely high quantum yield (close to 100%).

In practice, researchers can combine two semiconductor catalysts (powders, one for HER and another for OER) with a reversible donor/acceptor pair (mediator). The mechanistic picture for Z scheme water splitting is shown in Figure 2 - 60 and illustrated by the following reactions.

$$2H^+ \longrightarrow 2e^- + H_2 \text{ (Reduction of } H^+ \text{ to } H_2\text{)}$$
$$D + nh^+ \longrightarrow A \text{ (Oxidation of D to A)}$$
$$A + ne^- \longrightarrow D \text{ (Photoreduction of A to D)}$$
$$2OH^- + 4h^+ \longrightarrow O_2 + 4H^+ \text{ (Photooxidation of } OH^- \text{ to } O_2\text{)}$$

Reduction of protons occur by donation of electrons of CB and oxidation of an electron donor by VB holes, yielding an electron acceptor. The electron acceptor obtains electrons from VB and become again an electron donor. The holes generated in VB then withdraw electrons from H_2O/OH^- and accelerate water oxidation reactions.

Finding good water reduction/oxidation catalysts is important for development of Z scheme technique in tandem photo water splitting processes (Figure 2 - 60). The semiconductor catalysts work coordinately with redox shuttle mediators in aqueous electrolyte and promote surface reactions while restraining electron-hole recombination. In general, Z scheme technique can be classified into two types: photoelectrochemical device and photocatalytic device. In photoelectrochemical configurations, photoanode is coated with OER materials and photocathode is covered with HER species inside redox electrolyte solution. Therefore, such device consists of two electrodes for OER and HER reactions, respectively, and an electro mediator linking the two oppositely charged electrodes. With two electrodes suitable for each half reaction, the main advantage of tandem cell is that a larger fraction of solar energy can be utilized for water splitting reaction. Another advantage is that H_2 and O_2 evolution reactions can occur in spatially separated electrodes, allowing easy gas separation and reduce the overall cost.

Obviously, the lifetime for electron-hole pairs, conductivity and diffusion properties of those materials are still important to acquire good photoelectrochemical performances of tandem cells. Early successful cases using Z scheme included TiO_2—WO_3 in the presence of Fe^{3+} cation. In other cases, p-type semiconductor materials $SrTiO_3$ doped with Pt or Rh can be coupled with n-type $BiVO_4$ or Pt/WO_3 catalysts for Z scheme water splitting as well. IO_3^-/I^- redox pairs can also be used to transport photo induced electrons from CB of photoanode to VB of photocathode to facilitate electron transfer reaction.

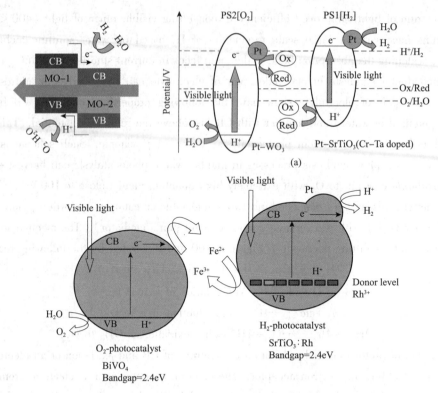

Figure 2 – 60　Z scheme technique of water splitting

Powdered photoelectrochemical catalysts have also been demonstrated to be remarkable performances in combining metal-complexes, graphene and semiconductors. As mentioned above, the Z scheme systems display good properties for producing H_2 and O_2 on different sites, while restraining the electron-hole recombination processes.

2.7.3　Catalyst development

(1) TiO_2 nanocatalysts

We have mentioned TiO_2 materials as effective semiconductor catalyst many times in previous sections. TiO_2 is often existing as white powdered form. Typical bandgap for TiO_2 is around 3.2 eV. Such band structure is suitable for absorbing UV spectrum in solar energy. Metal or non-metal doping has been widely studied to enhance the absorption width and better quantum yield. Doping non-metal elements such as N, C and S alters band structures by narrowing the bandgap. Metal ions can be employed enhance harvesting efficiency.

Black TiO_2. Generally, it is difficult for surface modified TiO_2 to cover IR spectrum, which accounts for 52% of solar irradiation. If the color of TiO_2 become dark, the coverage for IR range could be improved. Designing black TiO_2 can be achieved by hydrogenation of regular TiO_2 materials. High pressure H_2 treatment, or plasmonic treatment are among various accepted methods which have been studied in the past few years. Numerous studies have conducted for hydrogenation of pure TiO_2 crystals from ambient to high pressure (4MPa), with heating temperature around 400 ~ 500℃. This process often takes for several days. Nanowired TiO_2

materials display milder treatment conditions and shortened time (< 3 h). The bandgap for reduced TiO_2 is approximately 1.8 eV, covering most of IR region. Noble metal modified TiO_2 can display lowered treatment temperature. Because the strong H_2 spillover from Pt and Pd sites can facilitate TiO_2 reduction in the range of 160 ~ 500 ℃.

In addition to H_2 reduction, other chemical methods can also be used to modify TiO_2 materials. Ethanol, HF, $NaBH_4$ as well as metallic Al and Mg can be added when preparing TiO_2-based catalysts. Those chemicals and react with TiO_2 and obtain O species during reduction processes. The oxygen vacancies created during reduction significantly improved optical absorption in the visible and infrared region.

The reduction of Ti^{4+} into Ti^{3+} will induce another phenomenon, the defective structure. In other words, the original stoichiometric properties could be tuned due to the absence of oxygen species during surface reduction, to ensure electrostatic balance. More importantly, according to literature, the defects in the crystal play a critical role to improve the optical and catalytic properties of TiO_2. The defects actually disorder the original crystalline, changing the amounts of Ti—OH and Ti-H fragments. Previous books have summarized the general methodologies for reducing TiO_2 into black/dark colored TiO_2.

Clearly, characterization of black TiO_2 (Figure 2 - 61) is important to understand the structure-property relations. Electron spin resonance spectroscopy is very useful to study the existence of O vacancy. FTIR spectroscopy can also reveal the presence of Ti—OH group in the samples. XPS spectra and XRD patterns can also provide important information on the amount of Ti-H groups. In addition, theoretical studies also suggest valence band shift in black TiO_2 materials. The central focus for theoretical calculation is on the facet dependent adsorption/desorption of H_2 species on TiO_2 surface. The kinetic barriers for H_2 diffusion from surface to subsurface and combination with surface O containing groups have been determined for (101) surface by previous researchers.

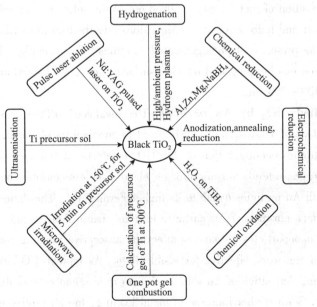

Figure 2 - 61 Preparation of black TiO_2

Surface modified TiO$_2$ by radiolysis and photo deposition. The most popular method to improve the photoelectrochemical performances of TiO$_2$ materials is to introducing metallic species onto the surface. For example, doping TiO$_2$ with Rh^{3+} and Bi^{3+} cations can expand the absorption spectrum. Au and Ag modified TiO$_2$ catalysts have attracted wide attention, owing to localized surface plasmon oscillations of CB electrons. Different studies have demonstrated that metal modified TiO$_2$ catalysts exhibit shifts in the Fermi level to more negative potentials. The size of metallic particles can affect shift of Fermi level. Photochemical and radiolytic methods are effective in synthesizing metal particles with controllable size on the surface of TiO$_2$.

Radiolysis method uses irradiation to excite electrons and radicals to reduce metal cations in a homogeneous medium. As a result, small and monodispersed nanoparticles can be obtained using this facile method. The main advantage for radiolysis method is to introduce homogeneous nucleation and growth in the whole volume of the sample. Various metal doped TiO$_2$ materials have been successfully prepared in solution. The fundamentals for radiolysis is the strong interaction of high energy radiation such as electron beams, X-rays with metal cations in the solution. The excitation and the ionization of solvent transforming water molecules into solvated electrons, H$_3$O$^+$, H·, HO· and H$_2$O$_2$ leads to strong reductive/oxidative effects. Solvated electrons and alcoholic radicals have strong reducing tendency and are able to reduce metal ions to lower valence states and eventually to metallic species. Hydroxyl radicals have very strong oxidation activities thus could prevent formation of metallic nucleation. Hydroxyl radical scavengers are often added to avoid this issue.

When solid support is in the aqueous solution, metal ions can diffusion from homogeneous medium into the confined space of porous materials. The penetration of ions inside solid support under radiation enable reduction of cations and formation of metallic clusters in the matrix.

Photo reduction of metal precursors on semiconductor materials is also proven to be effective. Photo irradiated deposition of metal ions is carried out in the solution containing metal precursors, semiconductor support and hole scavengers. Metal ions adsorbed on the surface of semiconductors can be reduced in the presence of photogenerated electrons. For example, Au nanoparticles with varied size distribution can be deposited on TiO$_2$ surface to ensure widened absorption spectrum of solar light towards 400~700 nm.

Surface modified TiO$_2$ by Au particles. The localized surface plasmonic oscillation of electrons on the surface of TiO$_2$ support can be used to enhance its photocatalytic activity. Before we discuss the electronic reconfiguration of Au/TiO$_2$ catalysts, a few facts should be taken into account: Rutile TiO$_2$ phase tends to show stronger electron-hole recombination rate. Anatase phase interacts strongly with Au particles owing to its higher Fermi level. The dielectric constant of TiO$_2$ could shift local surface plasmon of Au particles towards visible light region.

The size of Au nanoparticles is found to affect the absorptive range of solar energy. Small Au particles of 2~3 nm can work effective for visible light. As typical TiO$_2$ solids show absorption edge at 325~400 nm, deposition of Au with 3~13 nm in size can extend absorptive range to 560 nm (Figure 2-62). With further increase in metal loading, the absorptive edge can be enlarged to 650 nm. This plasmonic shift is due to the interaction between Au particles and TiO$_2$ as so-called Schottky barrier.

Charge carrier dynamics of Au/TiO$_2$ systems show that, electrons trapped by Au particles and injection into CB of TiO$_2$ under solar irradiation due to plasmonic response. Figure 2 – 63 shows that as wavelength increases from 365 nm to 560 nm, the excited electron voltage for pure TiO$_2$ support drops significantly from 100 mV to < 2 mV. In contrast, the dynamics for Au modified TiO$_2$ catalysts display steady performances under varied wavelengths.

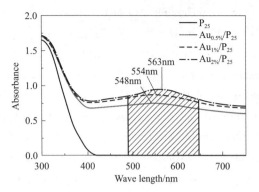

Figure 2 – 62   Surface modification of TiO$_2$ by Au particles

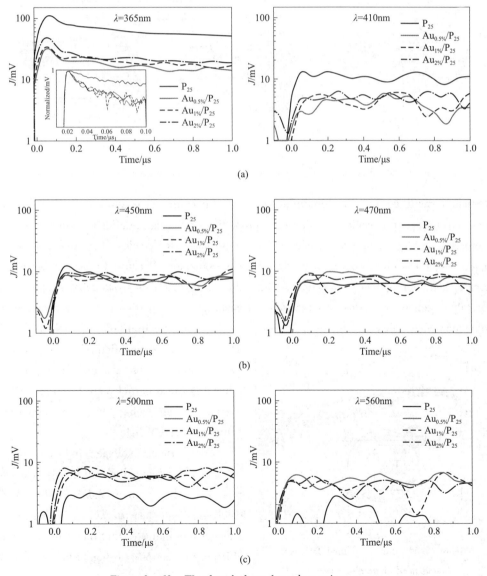

Figure 2 – 63   Wavelength dependent plasmonic response

Mechanisms for Au/TiO$_2$ catalyzed electron transfer reaction for solar light < 350 nm and above 450 nm are different. As shown in Figure 2-64, for solar light near UV region, electrons are excited from VB to CB of TiO$_2$. The hot electrons can then be transferred for reduction reactions on the surface of TiO$_2$ or Au particles. However, for solar irradiation close to visible spectrum, electrons in TiO$_2$ cannot be excited from VB to CB. Therefore, plasmonic response from Au trapped electrons is key for efficient photoelectrochemical water splitting. As shown in Figure 2-65, visible light irradiates on the surface of Au particles to induce electron transfer from Au to CB of TiO$_2$. Experimental studies have confirmed that trapped electrons at CB (Ti^{3+}) and holes (O$^-$) for electron transfer from Au particles to TiO$_2$ is the main mechanism for visible irradiation. Such behavior is particularly enhanced near the plasmonic wavelength for Au particles.

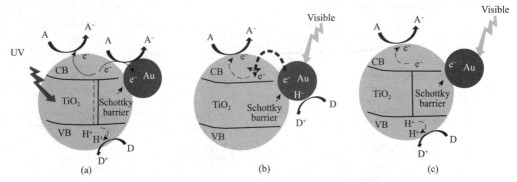

Figure 2-64 Mechanism of Au modified TiO$_2$ surface for enhanced electronic transfer

Surface modified TiO$_2$ by bimetallic particles. Bimetallic nanoparticles often display tunable surface, electronic, plasmonic and magnetic properties, differing significantly from monometallic counterparts, thus receiving extensive attention in the past few decades. For Au-based bimetallic nanoparticles, Cu is often known as good electronic modulator. Because Cu is also a well-known plasmonic metal for optical applications. Bimetallic AuCu/TiO$_2$ catalysts have shown enhanced photocatalytic activity for numerous reactions. Bimetallic AuCu particles show a decrease in the photoluminescence emission intensity, suggesting lowered electron-hole recombination rates. Therefore, AuCu modified TiO$_2$ catalysts exhibit better performances in charge separation as bimetallic AuCu particles are better electron sinks leading to an enhancement of photocatalytic activity under UV light.

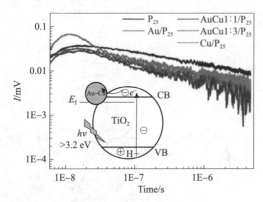

Figure 2-65 AuCu modified TiO$_2$ for enhanced photo absorption

The instant voltage induced by solar irradiation for bimetallic AuCu/TiO$_2$ catalysts is relatively higher than monometallic Au/TiO$_2$ catalysts, with fast decay in responses. This phenomenon indicates that bimetallic catalysts are efficient in electron scavenging than monometallic ones.

Different from bimetallic AuCu particles, the case for AuNi catalysts is established based on an alternative scenario. It seems that Au and Ni species form separated sites rather than alloy form. The quantum efficiency for AuNi/TiO$_2$ catalyst is enhanced to 20% ~ 40% for UV light (300 ~ 390 nm), which is much higher than Au/TiO$_2$ catalyst (<10%). Such catalyst display good performances for HER. The mechanism for the separated Au and Ni sited catalysts is proposed in Figure 2-66.

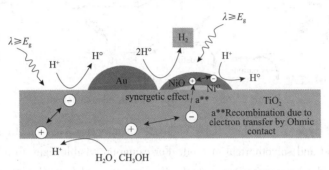

Figure 2-66 Electronic transfer mechanism on AuNi surface

In addition to Au-based bimetallic catalysts, other combinations such as AgCu and PdNi can also show good quantum efficiency. The wavelength dependence for PdNi/TiO$_2$ catalysts is analogous to that of AuNi/TiO$_2$ catalyst.

Synthesis strategies for TiO$_2$ heterojunctions. As mentioned in previous sections, heterojunctions are effective structures for enhancing photoelectrochemical water splitting performances, owing to narrowed bandgap of original semiconductors. The first type of heterojunction involves a material (A) with wide band gap which completely overlap the bandgap of another semiconductor (B). The excitation of electrons involves two parallel pathways, as shown in Figure 2-67. The second type of heterojunction involves partially overlapped bandgap structures. The CB of one material (A) is higher than that of another semiconductor (B). The third type of heterojunction is shown in Figure 2-67 (c). There is no overlap for two band structures. Other heterojunctions can also be utilized, including Z scheme, p-n and schottky types.

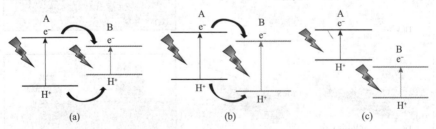

Figure 2-67 Typical TiO$_2$ heterojunctions

Students are familiar with Z scheme water splitting, however, Z scheme heterojunction avoid uses of homogeneous electron acceptors/donors. Electrons are transferred from CB of material B to VB of material A for consecutive excitation steps. The p-n junction is actually similar to type 2 structure (Figure 2-68). Solar irradiation induces electron transfer from p type semiconductor to

CB of n type material for reduction reactions, while hole carriers in n-type material migrates to p semiconductor for oxidation reactions. In other words, the junction formed by "n" and "p" materials facilitates the accumulation of photo generated electrons near n-type interface and the holes near p-type.

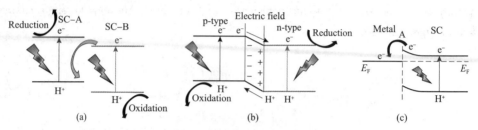

Figure 2-68 Three types of TiO_2 heterojunctions

Preparation of type 2 modified TiO_2 catalysts usually involves electrodeposition method, precipitation method and solvothermal method. For example, visible light absorbing Bi_2WO_6/TiO_2 film can be prepared by precipitation method. The formation of heterojunction is facilitated by a hydrophilicity-assisted dip coating method. Heterojunction of TiO_2-graphene interface was also prepared by facet controlled method. In particular, (101), (100) facets can be anchored on graphene by a hydrothermal method in the presence of various capping anions. During hydrothermal process, $Ti(OH)_4$-graphene hybrids were formed by introducting $TiCl_4$ precursors to aqueous HCl solution to ensure formation of (101) facet. Alternatively, (100) facet can be anchored with graphene using $Ti(OC_3H_9)_4$ as precursor in the presence of HF containing solution. Similar solvothermal method can also be applied for C_3N_4/TiO_2 heterojunctions.

MoS_2 catalysts hybridized with TiO_2 can be prepared *via* hydrothermal route. MoO_2 particles can be decorated on the surface of TiO_2 followed by sulfurization under elevated temperature to obtain MoS_2 phase (Figure 2-69).

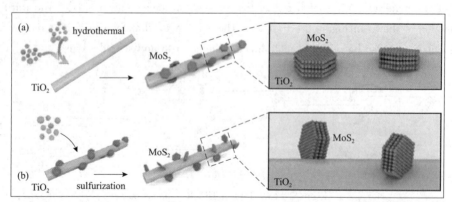

Figure 2-69 Surface modification of TiO_2 by MoS_2

Schottky heterojunctions are basically metal modified semiconductor TiO_2 catalysts. The performances have been discussed in the previous section. Preparation of such heterojunction structures actually involves basic impregnation method, or in-situ reduction method in the presence of light irradiation. For example, TiO_2 powders were dispersed in water-methanol solution containing metal chloride followed by UV exposure under argon for 90 min to reduce metal cations into metallic forms and deposited on TiO_2 surface. The consequent improved photoelectrochemical activity for water oxidation and reduction is also ascribed to combined plasmonic properties with Au, Ag and Cu metals.

Combination of p-type and n-type semiconductors can effective minimize the recombination of charge carriers after photogeneration. Several p-type semiconductor materials such as Cu_2O, Ag_2O can be combined with TiO_2. Interestingly, such combination may display both the characteristics of both Schottky and p-n heterojunctions, thus both photocatalytic activity and better charge mobility and separation can be well achieved in such hybridized systems.

Z scheme heterojunctions effectively reduce the use of homogeneous electron acceptors/donors. WO_3/TiO_2 materials were prepared by previous researchers to obtain Z scheme structures. Similarly, $TiO_2@NiS$ core-shell structures can be synthesized using solution method followed by annealing at elevated ($>500\,^\circ C$) conditions. For detailed preparation procedures, $Ni(OH)_2$ is often deposited on the surface of TiO_2 materials before sulfurization step. The thickness and morphological features of NiS species can be controlled by the extent of sulfurization such as temperature, pressure and processing time.

(2) **Vanadate materials**

$BiVO_4$ is the most popular candidate for OER applications. The optical properties of $BiVO_4$ is very similar to that of WO_3 and most suitable for OER application. Combining $BiVO_4$ with WO_3 of good charge transfer properties with layered structures can be promising options for heterojunctions.

To compress recombination of surface charges, $BiVO_4$ loaded with CoO or NiO is found to be better. The stability could also be improved in terms of pH and electrochemical potential in aqueous medium, after incorporation of a micro or nanstructured $BiVO_4$ rather than bulk materials.

Replacing Bi with Ag leads to the formation of perovskite structures $AgVO_3$ with 2.3~2.5 eV band gap. Similarly, Ta and Nb can also form perovskite structures with 2.4~2.8 eV band gap, where hybridization of O p and Nb/Ta d orbitals is key for water splitting reactions. Such hybridization also enables formation of octahedral NbO_6 phase.

(3) **Hybridized Fe_2O_3**

Iron oxide materials are readily and abundantly available, although iron ions can exist in various forms. Commonly iron oxides are in red-brown color, coupled with non-toxic properties, showing good potential in large scale implementation.

Hematite Fe_2O_3 is the most common form of iron oxide. The hematite structure is based on close packing of anionic O^{2-}, which is arranged in a hcp lattice. Such arrangement produces pars of FeO_6^{9-} octahedral with edges of three neighboring octahedral in the same plane and one facing an octahedron in an adjacent plane. Oxygen anions and high-spin (d^5 configuration) iron cations affects the orientation of iron atoms spin magnetic moment thus the observed bulk magnetic properties can be influenced. The trigonal distortion of FeO_6^{9-} octahedral causes the spins to become slightly canted leading to destabilization of their perfectly antiparallel arrangement.

Markovic and colleagues reported that, OH—M^{n+} (M: Fe, Co, Ni, Mn) bond strength is critical for HER and OER activity. This relationship exhibits trends in reactivity (Mn < Fe < Co < Ni), which is governed by the strength of the OH—M^{n+} energetic (Ni < Co < Fe < Mn).

Trudel and Berlinguette validated an effective photo deposition method for preparing amorphous metal oxide for OER application.[2] Iron oxide materials synthesized used this method show superior activity compared with conventional hematite materials. XRD characterization confirmed the formation of α-$Fe_{100-y-z}Co_yNi_zO_x$ phase, which is key for enhanced performances for water oxidation.

In another study, Bard and Mullins studied amorphous FeOOH materials for OER application. They found that the doping of Co-borate to FeOOH led to synergy for light absorption and improved conductivity. The effect of doping also involves alteration of surface energy. For example, adding Fe^{3+} to NiO crystals leads to predominate surface coating thus agglomeration of NiO clusters. Such unique surface-bulk composition induces formation of Ni—Fe spinel phase.

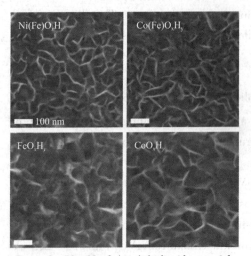

Figure 2-70 Metal (oxy) hydroxide materials

Reduction of Fe^{3+} and Ni^{2+} precursors using electrochemical method could also lead to the formation of $FeNiO_x$ thin film. Surface morphologies are different under varied pH conditions. The unique preparation method produces transparent binary oxide film thus show good potential for absorbing and penetrating solar irradiation with different wavelengths. It is important to mention that, the interface among different phases, for example, $Ni(OH)_x$/NiOOH hybrids, are held together with only by noncovalent bonds (Figure 2-70). The structures are often disordered, allowing for electrolyte access within the catalyst bulk and inter-sheet space.

In addition to mixed metal oxides, iron oxide materials can be manipulated to coordinate with other oxide materials to promote electron transfer and lower the significance of mass transfer limitation. For example, $FeOOH/CeO_2$ tubes can be grown under solvothermal conditions.[7] The lattice match between CeO_2 (044) facet with FeOOH (310) and (211) surfaces lead to heterolayered structures. XPS characterization has confirmed that, O in Ce—O negatively shifts approximately 0.15 eV, while Fe—O—Fe positively shifts to 0.19 eV. The altered binding energies at interface promotes electron transfer during OER application thus intrinsic kinetics can be improved under mild condition. Combining FeOOH with conventional $BiVO_4$ catalysts could improve stability by increasing electron transfer rate during formation of O—O bond.

In addition to oxide materials, metal nitrides could also mitigate the performances of OER application. Co_3FeN_x structures also display good performances for OER.[8] Surface characterization confirm that coexistence of Co^{3+}—O/N bonds and Co—OH bond could be dual functional sites for the formation of O—O bond. Similar hybridization effect can also be seen in Fe_3O_4 decorated CoS_x system. The dual cationic or anionic effect contributes to tunable metal—O bonding strength thus enhancing formation of O—O bond. Doping N into solid support also favor electron transfer from metallic particles to reactants. Doping other promoters such as P and Se could lead to the generation of amorphous structures, characterization show that metal—P or metal—Se bond can be formed. Phosphates are reported to be good proton transport mediators at catalyst surface. $Fe(PO_3)_2$ species with inequivalent P—O in P—O—P bond is attributed to remarkable performances. More importantly, metal—OH/metal—OOH can be easily tuned at low potentials thus formation of O—O bond can be accelerated in such system.

Low dimensional materials can be coupled with defected graphene materials to exploit their remarkable properties for electron transfer. For example, Ni—Fe based layered hydroxide materials have been electronically heterassembled with negatively charged graphene. TEM and XRD characterization suggests that such electrostatic interaction induces delamination of MO_6 octahedral structures from 2D network (Figures 2–71, 2–72). As a result, the redox circle for Ni^{2+}/Ni^{3+} pairs can be facilitated.

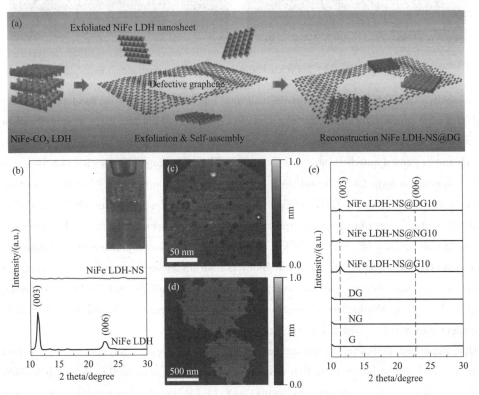

Figure 2–71 Ni—Fe layered hydroxide coupled with defected grahene

Co-based catalysts are regarded as important OER materials with cost-effective advantages over Ru and Ir catalysts. For most current studies, the primary focus is still on the following aspects: engineering active sites and electronic conductivity. Introduction of oxygen vacancy, doping with heteroatoms and coupling with other conductive materials may be effective in tailoring

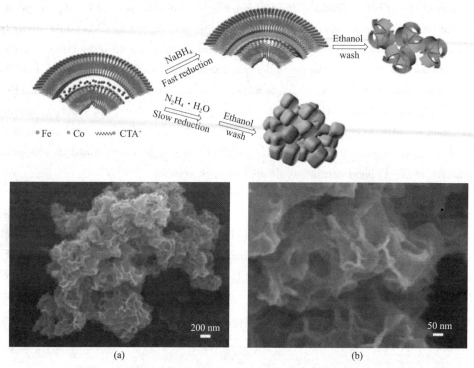

Figure 2-72 FeCo nanosheet with defected surface structures

such properties. Synthesis of 2D Co oxide materials has been found to be most popular in recent decades. Facile phase transformation from CoO to Co_3O_4 nanosheet results in atomically dispersion of active sites and high current density compared with commercial Pt/C catalysts. Fabrication of large amounts of coordinated-unsaturated Co atoms. The unsaturated Co species may be exposed to oxidative species to form Co^{3+}, which is active site for OER application.[16] Electrocharacterization has shown that, oxidation of Co^{3+} to Co^{4+} for electron transfer reaction. DFT calculation also suggests that, Co^{4+} contributes to adsorption of nucleophilic species to form OOH^- intermediate, inducing lattice vacancy on catalyst surface. In addition, the thin film structures facilitate diffusion of molecular O_2 thus surface utilization efficiency is high under high current density conditions. In addition, the incorporation of a second component such as Ni and Fe could induce the formation of Fe—O or Ni—O by sharing some oxygen species with Co species. In other words, the local environment can also be polarized in such amorphous structures.

One important technique is cation exchange. The difference between original cation and foreign species leads to lattice distortion at local environment. For example, Bi^{3+} can replace Na^+ and Ta^{5+} cations thus band-gap narrowing can be achieved. Importantly, the doping of heterocations into original structures could also change the crystal structures into another form. A typical example of $NaLaTaCrO_x$ composites, where the stoichiometry number of x, due to varied cationic compositions, causes crystallinity change from monoclinic to orthorhombic structure. Among many candidates, $BaTa_2O_6$ catalyst in orthorhombic phase show very good performances. If doped with Fe_2O_3 and TiO_2 particles with layered structures, the performances could be further and dramatically improved as a result.

Vanadate photocatalysts with monoclinic metal oxide structure (e.g., $BiVO_4$) are better than conventional WO_3 materials for OER performances. But the poor stability limits the application of such material.

Low dimensional materials also display unique electron transfer properties due to surface defects. Generation of amorphous layer facilitates electron and proton transports, where Co^{4+} species are key for redox processes. Experimental studies demonstrated that, amorphous $CoO_x(OH)_y$ layers enhance O_2 diffusion. This is because tunable Co^{4+}/Co^{3+} ratio is achievable in alkaline medium. Identifying the intrinsic structures of Co sites has been paid on large attention, with yet substantial progresses. But surface characterization of Co catalysts indicates that the edge-sharing CoO_6 octahedra clusters actually exist in all thin film structures. The chemistry of O—O bond formation probably proceeds at periphery of Co—O clusters. Doping heteroatoms obviously will induce structural distortion, due to the fact that heteroatoms have different atomic radius compared with the existing cations/anions (Figure 2-73). Such structural disturbance causes charge redistribution and variation of bonding energy of surface oxygen.

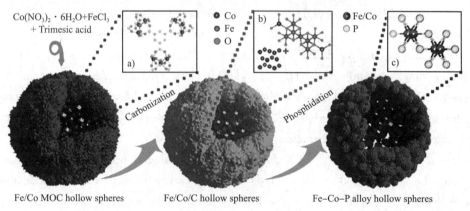

Figure 2-73　FeCo-P alloy catalyst

(4) Hybridized $MTaO_3$ materials

Tantalate perovskites are well-known for very good overall water-splitting reaction under UV irradiation. $NaTiO_3$ catalysts loaded with co-catalysts such as NiO can effectively achieve more effective water splitting reactions. Bi-doped $NaTiO_3$ display a narrowed band-gap structure with good visible light absorption > 390 nm. The Mott-Schottky analysis reveals that, the flat band is more negative after doped with Bi cation. Similarly, co-doping mixed species La—Ni, La—Co and La—Fe in such material could also enhance absorption for visible light. This is because that, crystal structures can be transformed from nonoclinic to orthorhombic with modulation of chemical composition.

Experimental studies also confirmed that, doping both anion and cation in $NaTaO_3$ could be useful for enhanced visible absorption. Computational studies showed that, N, F, P, Cl and S doping are all effective. Replacing Na with other alkaline and alkaline earth metal cations also leads to more exciting results. $BaTa_2O_3$ materials in orthorhombic phase was found to be most active among other cationic candidates such as K, Mg and Li.

(5) Carbon-based materials

Incorporating carbon materials with other semiconductors can effectively promote charge carrier separation. Carbon can play as a co-catalyst for transporting electrons from semiconductor materials. Carbon and metal oxide can form type 2 heterojunction structures. Photo induced electrons can be excited to CB of metal oxide, after which electron transfer across the interface can facilitate HER reactions on the surface of carbon materials, which are often deposed with catalytically active species for reduction reactions. Once solar light irradiated on photocatalysts excites electrons to CB, after which charge separation occurs. Electrons flow across the heterojunction to carbon surface, where catalysts for H_2 evolution functionalize and generate molecular H_2. Minority hole carrier can also migrate to carbon and facilitate O_2 evolution reaction.

(6) Metal chalcogenides

Metal chalcogenide materials have been extensively investigated for electrochemical catalysis. The anisotropic nature of metal chalcogenides, which are direction—dependent to confine motion of electrons, holes, excitons, phonons and plasmons in surface catalysis. Various metal chalcogenides, such as Cu_2S, FeS_2, CoS_2, NiS_2, TiS_2, etc. are among mostly investigated subjects.

Morphology affect electronic configuration in metal chalcogenide materials. For example, single-layered and multi-layered MoS_2 materials have distinct bandgap structures with different spin-orbit spinning. Construction of 1D, 2D and 3D metal chalcogenides is the main focus in numerous research papers and patents. In general, four methodologies have been applied to manufacturing photo active materials, including intrinsic growth, shape-guiding agent growth, oriented attachment and chemical transformation.

In intrinsic growth process, the surface energy of basal and prism planes determines the eventual morphology. Hydrothermal synthesis in the presence of polymers is among the most popular methodologies. According to Hume-Rothery rule for solid solution, metallic elements with insignificant radius difference ($<15\%$), similar crystal structures, valence and electronegativity are essential criteria. SeTe alloy, Bi/Sb—S/Se/Te with highly anisotropic properties have been synthesized by previous researchers. Overall, experimental parameters during intrinsic growth processes, together with polymer agents affect the specific surface energy thus leading to 1D or 2D morphologies.

Shape-guiding growth occurs in the presence of selected surface modifying agents. To obtain CdSe crystals, phosphonic acid, phosphine oxide, trimethylamine and carboxylic acids can be introduced in solution based methods. The key principle to design metal chalcogenides with selecting appropriate agents with desirable binding strength with specific facets.

Chemical transformation from a premature crystal is a convenient method. Generally, three types of chemical transformation can be applied, alloy formation, ion exchange and galvanic replacement. Alloy formation method can be used to prepare alloy structures by mutual diffusion between atoms in premade solids and dissolved atoms in reduced form. Diverse morphologies such as particles, tubes and wires can be accomplished for ZnS, CdS, PbS, Ag_2Se, Bi_2Te_3, CoTe materials.

Finding active and durable metal sulfide materials for photocatalytic overall water splitting is very difficult, as most materials are poorly resistant to hole oxidation. Compared with metal oxides, sulfides often have narrow bandgap structures because S element has lower electronegativity than O. In most cases, sacrificing agents are necessary to enable durable performances of metal sulfides.

Incorporation of sulfide into lattice of metal oxides forming oxysulfides is expected to improve stability of semiconductor catalysts against hole oxidation. Hybridization of S 3p and O 2p leads to stabilized structures. For example, $Sm_2Ti_2S_2O_5$ and $Ln_2Ti_2S_2O_5$ are found to show good durability for water splitting reaction. $Y_2Ti_2S_2O_5$ is found to have very narrow bandgap of 1.9 eV corresponding to absorption edge of 650 nm of solar energy. Studies found that maximum level of VB is shifted owing to 3d of Y, contributing to narrowed bandgap.

Existing studies on metal sulfides have been covering the synthesis, characterization.

Summary

Students should understand the basics for half reaction for water splitting, and photocatalytic mechanism for electron transfer during reduction and oxidation reactions. Students with engineering background should also pay attention to the reactor design and process development, from manufacture of solar panel, photocatalytic reactors as well as gas separation/purification units. Nanostructured materials are key for improved solar-to-chemical energy conversion. Students should be able to apply fundamentals of material chemistry to rationale the design of novel catalytic materials for complete water splitting, HER and OER reactions.

Material design is the key for the second part of this chapter. Students should be able apply the basic wet chemistry knowledge for electron-hole creation, migration and separation.

Exercises for Chapter 2

(1) Plot the solar energy conversion route to fuel and industrial chemicals.

(2) Configuration of electro, photo and photoelectron water splitting devices.

(3) Solar energy spectrum and utilization technique.

(4) Mechanism and elementary steps for photoexcitation.

(5) Definition of band gap and typical band gaps for semiconductors.

(6) Surface reactions involved in photo conversion of water to H_2 and O_2 (electronic transfer reactions).

(7) Band structures of electrolyte-electrode interface.

(8) Nano effect on band gap configuration.

(9) Plot two semiconductor and semiconductor-metal junctions, discuss the mechanism.

(10) Discuss your understanding on how doping hetero elements into semiconductors affect band structures. Give a few examples.

Chapter 03 Biomass Conversion to Fuels and Chemicals

3.1 Introduction

The 21st session of the Conference of the Parties (COP21) of the United Nations Framework Convention on Climate Change (UNFCCC), which was held in Paris during November to December, 2015, include the objectives to peak greenhouse gas emission as soon as possible. This goal is to limit the temperature rise on average above pre-industrial level to well below 2℃. Biden administration in newly elected government has announced to pursue an ambitious 1.5℃.

Biofuels and biochemical clearly represent the most promising renewable carbon resources on our planet. Global production for biofuels has increased 10 billion L in 2018 to reach a record 154 billion L. This volume has doubled the growth of 2017. The United States and Brazil were the largest biofuel producers, with Asian countries as potential contributors for bioethanol and biodiesel production.

Over the span of 2019—2024, biofuel output is expected to increase by 25% to reach 190 billion L (Table 3-1). China has become the third largest contributor in biofuel production with better market prospects across the globe.

Table 3-1 Global production on biofuel

Biofuels	2018 Growth/L	2019 Output/L	2019—2024 Increase	2024 Output
Ethanol	6.6	110.4	19%	130.3
Diesel and HVO[a]	3.6	42.6	34%	57.1
Advanced fuel	0.2	1.4	100%	2.8

[a] HVO = hydrotreated vegetable oil.

However, owing to the high oxygen content existing in biomass feedstocks, bio-oxygenates might not be the perfect candidates for fuel production. Conversion of biomass resources, such as waste fruit/food, human waste from cities, non-edible oxygenates including cellulose, hemicellulose and lignin to generate heat and bio-fuels have attracted extensive attention.

In this chapter, the history of biomass utilization for fuels, fundamentals for biomass conversion to fuels and chemicals will be discussed with most recent research results in both academia and industry.

3.2 Biomass: the ultimate renewable carbon resource

3.2.1 Source of biomass

Conversion of biomass to fuels should consider the use of landscape in growing biomass resources. Potential sources of biomass should be evaluated for both energy and economic basis. Comprehensive balance should be considered for all potential energy and financial input during cultivation, harvesting, transportation and processing of plant materials. Emission of CO_2 and other greenhouse gases such as NO_x during application of fertilizers should also be taken into account.

Today's production and manufacture of biofuel products are heavily dependent on using of vegetable oils, animal fats, terpenes and sugar or starch crops (Table 3 - 2). Competition between food consumption and fuel demand has caused both technological and economic disturbance in many countries. The situation could be worse if increasing demand of sustainable and low-carbon fuels continue compete for land and food uses, particularly in developing countries.

Table 3 - 2 Production of biofuels and mileage information

Biofuel	Natural resource	Cruising range/(km/ha)
Biogas	Maize silage, *etc.*	67,600
Biomass to liquid	Straw	64,000
Rapeseed oil	Rapeseed	23,300
Biodiesel	Rapeseed	23,300
Bioethanol	Corn	22,400

Note: Car mileage: gasoline, 7.4 L/100 km; diesel, 6.1 L/100 km.

Two types of biofuel have been largely implemented as drop-in alternatives in industry. The most prominent examples are bioethanol and biodiesel. Both of the biofuels have undergone two-generation development in the past two decades.

3.2.2 Bioethanol as sustainable fuel

Bioethanol is not a new concept for sustainable biofuels. Production of bioethanol from biomass has been largely implemented in North and South America, *e.g.* United States, Brazil, *etc.* Bioethanol is nontoxic and can be widely used as oxygenated fuel additives to replace traditional ones such as MTBE, which is hazardous to environment. The kay advantage for bioethanol is that, automobiles can utilize fuel blends consisting of 15% bioethanol without modification. At present, there are more than 10 million flexible fuel vehicles in North America. It has been demonstrated that bioethanol is compatible with current automobile technologies and fuel distribution networks. However, environmental studies also present the potential hazards posed by using bioethanol, such as follow:

(1) Production

Typical synthesis of bioethanol requires sequential procedures including biomass-to-sugar conversion, fermentation of sugars to ethanol and purification of ethanol from fermentation mixture. It is interesting to find that the overall energy requirement for bioethanol process is actually very high for systems based on sugarcane and starch. In Brazil there have been numerous arguably successful demonstration as capacity of bioethanol can partially substitute gasoline, as recognized by Brazilian government in the 1970s. Brazilian bioethanol became commercially competitive since 2007, as the capacity was 16 billion L requiring approximately 5.6 million hectares of agricultural land. But this only account for 1% of agricultural land in Brazil. US production of bioethanol has reached 1.2 billion L in 2006. But statistic analysis shows that, the energy yield associate with bioethanol production from corn is only 1.3. Therefore, it is clear that the energy obtained from conventional bioethanol is minimal and consumes large quantities of extra energy input.

The second generation of bioethanol is particularly attractive as lignocellulosic biomass is relatively cheap, widely available and abundant, compared with corn and food-based resources. Use of non-food biomass could provide renewable fuels without compromising food supplies. Life cycle analysis demonstrates that greenhouse gas emission could be reduced up to 50%~80% using non-food based biomass for bioethanol production. Economic saving compared with core-derived bioethanol is approximately 25%~40%. But it should be noted that additional unit operations are needed to break lignocellulose polymers into sugars. A variety of methods have been applied to address this problem, including milling, pyrolysis in the presence of acids, steam explosion, ammonia fiber explosion, ozonolysis, acid/base hydrolysis, and oxidative delignification. Organism such as fungi can also be used to decompose waste cellulosic materials into fermentable sugars.

A second problem associating with lignocellulose-derived bioethanol is separation of C_5 and C_6 sugar mixtures. Microorganism digests the sugar mixtures at much slower rate owing to the existence of pentoses. Those two problems have posed technological challenges for commercialization of second-generation bioethanol.

(2) Technological difficulties and R&D schemes

Cellulose, hemicellulose and lignin need to be decomposed for monomers before further processing (Figure 3-1). For waste biomass such as straw and wood, lignin materials shield cellulose from attack by enzymes and other chemical agents. In biological and enzymatic processing, chemical treatment followed by enzymatic breaking of plant fibers are often involved.

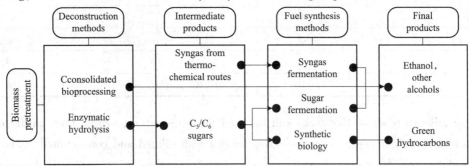

Figure 3-1 Biological production scheme of bio-based fuels and chemicals

The main purpose for pretreatment is to separate cellulose and hemicellulose from lignin, making both accessible to enzymes. The efficiency for sugar releasing by enzymes is only 20% without chemical treatment of lignocellulose, while the number goes up to 90% after treatment. In addition, since cellulose is often highly crystallized and water insoluble, decrystallization is necessary for enzymatic digestion.

Followed by chemical treatment including acidic, hydrothermal, alkaline or stream explosion, separation of cellulose and hemicellulose from the mixture is required for further processing with aqueous solution containing enzymes, to further formulate C_5 and C_6 sugars. This step is often referred as hydrolysis or saccharification. Finally, the fermentable sugars are fed into a solution containing microbes for biological conversion to ethanol. Through genetic engineering, microbes can also conversion sugars into higher alcohols and even hydrocarbon molecules which could be blended with fossil-derived gasoline, diesel and jet fuels. Key technological barriers for biochemical synthesis of ethanol are summarized as follow:

(a) Toxicity of pretreatment by-products and fuel synthesis products to the microbes used for synthesizing fuel. (b) Inability of existing microbes to process all types of sugar in lignocellulose efficiently. (c) Lack of biological processes producing hydrocarbons. (d) Lack of valuable co-products.

(3) Biochemical and biological conversion routes

Since approximately 1.3 billion tons of cellulosic biomass are available, it would potentially replace 20% of U.S. petroleum consumption. The first generation of bioethanol is derived from corn stover and food resources. Other feedstocks include switchgrass and short-rotation, hard woody materials such as poplars. Although composition varies differently, they are broken into mainly C_5 and C_6 sugars. The typical composition for corn, switchgrass and poplars are presented as below (Table 3-3).

Table 3-3 Composition of grain and cellulosic feedstocks

	Corn grain	Corn stover	Switchgrass	Poplar
Starch	72~73	Trace	Trace	0
Cellulose/hemicellulose	10~12	63~74	60	73
Lignin	0	14~18	10	21
Other sugars	1~2	3~5	6	3
Protein	8~10	1~3	5	0
Oil/other extractives	4~5	2	13	3
Ash	1~2	6~8	6	0.5
Total	96~104	90~110	100	100

Basically two types of feedstocks can be used for fermentation, syngas from gasification of biomass and hydrolysis sugars. Commercial processes with diluted and concentrated acids, have been used for over 50 years. Actually no commercial processes using enzymatic hydrolysis of cellulose exist in industry, although it is believed that this method could provide long-term solutions for reducing ethanol production cost.

In general, cell wall polysaccharides are more difficult to break down into small species compared with carbohydrates. Research results on solid pretreatment has been demonstrated to be beneficial for improved sugar productivity. Figure 3 - 2 presents the cost analysis for dry mill and direct enzymatic hydrolysis.

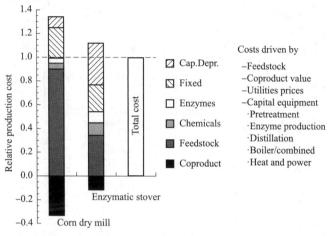

Figure 3 - 2 Cost analysis for bioethanol production

As a secondary technological barrier, pentosane utilization and poisoning need to be addressed. This is because pentosans are not metabolized through glycolysis in cells. Although there are some organisms can metabolize pentosans. But such organisms only show half of the reaction rate for common glucose fermentation processes.

The third barrier is process development. The diversity of feedstocks is a key limiting factor which restrain reliable scale-up design for fermentation processes. Three aspects should be considered and relevant issues are urgently needed to be addressed. Sustainable feedstock supply systems must be developed. Processes must be proved at scale. Societal and environmental benefits must be rigorously validated.

Fermentation of syngas is another important synthetic route for bioethanol. Ethanol can be produced by Butyribacterium methylotrophicum and at least six acetogenic clostridia (Figure 3 - 3). The reported optimal productivity of ethanol is approximately 48 g/L and 3.1mm/h. In general, acetate is the main product from syngas fermentation. By varying the culture conditions like pH and gas composition, it is possible to produce ethanol as the main product from syngas fermentation. Studies have shown that, *C. ljungdahlii* favors the production of acetate at a pH of 5 ~ 7. But ethanol is the dominant product at a pH of 4 ~ 4.5. Some companies already run pilot plants with a capacity of 15,000 gallon per year using steel mill off gas. The pre-commercial plant in Taiwan was operated with 37,900 L per year.

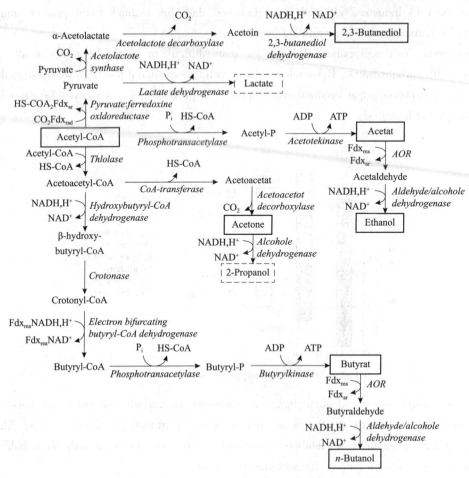

Figure 3-3　Pathways to possible products of syngas fermentation from wild type and engineered acetogens

3.2.3　Biodiesel as sustainable fuel

(1) The motivation

Biodiesel is considered as one of the most greatest options for sustainable fuels. This is because that biodiesel is drop-in ready for heavy duty trucks in many typical applications. Department of Energy in U.S. have already estimated an over 10% of domestic market by the end of 2020. However, this target indicates a six billion gallon productivity per year, which is a grand challenge on feedstock basis. Despite a vast market in E.U. with a 6% replacement early in 2005. The target for 10% market is very difficult to achieve.

Biodiesel is very versatile, meeting niche market requirement for combustion properties. The most outstanding properties of biodiesel include good lubricity, low sulfur content and good combustion performance. Biodiesel can be used in national parks, ports and major waterways owing to its environmental benign features. Pure biodiesel can be blended with petroleum diesel for many uses including military, transportation, boats. Assessment shows that pure biodiesel can reduce roughly 78% CO_2 emission compared with petroleum diesel. Biodiesel also leads to lower particulate matter pollution and SO_x emission.

Another reason that biodiesel market has grown rapidly is the existing legal infrastructures. Although the prices vary slightly based on end applications, biodiesel products are actually discounted driving consumers' acceptance. Different from bioethanol, logistical complications actually do not exist in this area. For example, transporting ethanol requires contract which should be documented in advance. However, biodiesel can be transported in current pipelines at a much lower cost of a few cents per gallon, well below that of shipping by road or naval transportation. The most important feature is that, biodiesel has drop-in features and can be blended easily with petroleum-derived products with comparable heating values.

Using biodiesel can also cause some problems. The heat duty is slightly lower than petroleum-derived ones. Although SO_x emission is lower, NO_x might be a potential issue. Biodiesel is biodegradable. But it also indicates that the durability is not as good as conventional diesel, owing to the presence of large amounts of oxygen-containing groups. Storage of biodiesel may face additional problems. Because biodiesel can degrade plastic and natural rubber gaskets and hoses.

The existing challenges for biodiesel industry is the difficulty in standardizing the quality assurance across various production facilities. It is particularly difficult to set up the standards as biodiesel can be obtained from different types of vegetable oils and some animal fats. Properties of each type of oil vary such as flow characteristics and co-flow features. Industries are still seeking technological methods to establish reliable quality measuring standards for biodiesels regardless of feedstocks.

(2) Production method

The first generation of is produced *via* transesterification of vegetable oils. Crude vegetable oils cannot be utilized as biodiesel due to high viscosity. However, transesterification with lower alcohols such as methanol and ethanol yields fatty acid esters in oil phase and glycerol as the by-product in aqueous phase. This process demands addition of acid or base homogeneous catalytic materials to facilitate the formation of fatty acid esters. Transesterification reactions are reversible, where the equilibrium is affected by reaction conditions, such as feedstock compositions, temperatures and pH values. When homogeneous alkaline species are used as catalysts, attention should be paid to ensure minimum soaps are formed during production. This is because soaps can easily cause emulsion and leading to severe separation problems during processing. Despite several potential issues, homogeneous NaOH and KOH catalysts are routinely used instead of acidic ones, as basic conditions could facilitate faster transesterification reactions.

The source of vegetable oil has a considerable impact on the final quality of biodiesel. For example, biodiesels obtained from palm oil and soybeans exhibit very different viscosities, cloud points, and combustion properties, making these fuels suitable for various applications. The economics are also influenced by the source of vegetable oils. It is estimated that, the cost of vegetable oil could account for 70%~90% of total operation costs. According to industrial data, the cost of biodiesel production in E. U. using Malaysian palm oil is $0.8 ~ 1.2 per gallon equivalent. One way to improve the economics of biodiesel production is to valorize the by-product, glycerol. Sales of glycerol has been found to marginally improve the cost-effectiveness of biodiesel production processes. Transformation of glycerol to other value-added products could further enhance the atom efficiency of the biodiesel industry.

In general, production of biodiesel consists of five stages, treatment of raw materials, alcohol-catalyst mixing, chemical reaction, separation and purification of products (Figure 3 – 4). For pretreatment of raw materials, it is advisable that basic catalysts might be leading to forming soaps in the downstream sections. The efficiency of reaction diminishes with increasing content of fatty acid in the vegetable oil. Use of basic catalyst is viable if the acidic content is below 2%. In the cases of highly acidic raw materials derived from animal fat, poultry, pork, vegetable oil from cotton, coconut, acid catalysts are more viable for preliminary transesterification to reduce acidity. Overall, acidity level (<0.1 mg KOH/g), humidity (<500 ppm), peroxide index (<10 meg/kg) and non-saponificable substances (1%) should be considered for treatment of raw materials.

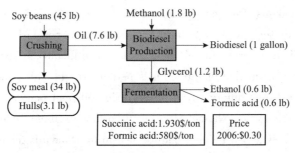

Figure 3 – 4 Production scheme for bio-diesel, bio-alcohols and acids

Catalysts and alcohols need to be completely mixed prior to reactor section. It should be noted that alcohols must be water-free to ensure good efficiency of reaction. NaOH and KOH are among the best basic homogeneous catalyst materials. For industrial scales, sodium or potassium methoxide or methylates are also available commercially. Acid catalysts including sulfuric acid, sulfonic acids and hydrochloric acid. But relevant studies are very limited. Heterogeneous catalysts that have been considered for biodiesel production include enzymes, titanium silicates, and compounds from alkaline earth metals, anion exchange resins, *etc*. For solid catalysts, we often need relatively higher reaction temperature to facilitate transesterification reactions, partially owning to low surface area of metal oxide species. In addition, solid base/acid catalysts are often easily saturated with polar molecules such as water and glycerol. Therefore, hydrophobic surface is preferable for practical operation. Examples of materials currently being considered for biodiesel production include basic zeolites, hydrotalcites, metal oxides, insoluble basic salts, immobilized organic bases, supported basic oxides, and alkali earth oxides.

Employment of solid acid catalysts could mitigate the formation of soap faced in homogeneous alkaline catalysts. However, acid catalyst may partially promote first and second transesterification, therefore, vegetable oil cannot completely convert into biodiesel products. But use of solid catalytic materials undoubtedly facilitate more efficient continuous process, leading to an approximately 40%~50% reduction in capital costs and 30%~60% energy saving. Therefore, application of solid catalyst would be the main focus of future diesel production.

Alcohol/oil volume ratio is an important process parameter. The stoichiometry of reaction is 3 mol of alcohol reacting with 1 mol of fatty acid ester to generate 3 mol of fatty acid methyl ester and 1 mol of glycerol. The general reaction scheme is described as follow:

$$RCOOR' + R''OH \leftrightarrow R'OH + RCOOR''$$

RCOOR' represents the raw ester as starting material, while R''OH is the alcohol. RCOOR'' is the resultant fatty acid ester while R'OH is another type of alcohol (glycerol in this case). If methanol is used as the starting alcohol, then the product will be fatty acid methyl ester.

Since this reaction is reversible, excess alcohol/oil ratio is necessary to ensure equilibrium is shifted towards fatty acid methyl ester side. Although high alcohol/oil ratio would not modify the properties of final products, it will make separation of biodiesel from the mixture difficult. An alcohol/oil ratio of 6 is often employed in practical operation.

As already mentioned, transesterification reaction is reversible, which actually involves three important steps. Triglyceride (vegetable oil) reacts with one methanol molecule to form diglyceride and one fatty acid methyl ester. Consecutive reactions of diglyceride and methanol eventually produce glycerol and additional two molecules of fatty acid methyl esters as final products.

There are several technological and economic disadvantages for first generation of biodiesel production. Use of homogeneous base and acid catalysts has caused equipment corrosion problems. Transesterification reactions also require an excess of alcohol and more suitable for batch operation rather than continuous production. Soluble catalysts can also migrate into glycerol phase and leads to loss of catalyst species and contamination of downstream products. The homogeneous alkaline species cause saponification. Solid base/acid catalysts are relatively easier to separate from product mixture, but poor durability is still an issue. But overall solid catalysts can selectively promote transesterification reaction and are tolerant to impurities in feedstocks.

Separation of reaction mixture can be achieved by decantation. The mixture of fatty acid methyl esters can be separated from glycerol phase because they have different densities. Two phases can form immediately as the stirring is stopped. Catalyst and excess alcohols will concentrate in glycerol phase (higher density), while unreacted triglyceride, diglyceride and monoglyceride are concentrated in upper phase.

The unreacted raw materials can be detrimental to the quality of methyl esters. Because it may undesirably increase cloud point and pour point. Therefore, the product should be washed and purified to meet the established standard of quality. The first washing step is using acidic water to remove methanol, catalyst and glycerol as well as other water soluble components. Two additional steps using water are also needed. Eventually, water components will be eliminated by drying, after which the purified product is ready for further testing as biodiesel.

Glycerol also needs purification as biodiesel plant only produce crude glycerol with poor quality. Distillation is used for glycerol purification and recovery of methanol. The cost-effectiveness depends on the scale of production. The steady growth of biodiesel market has also stimulated use of glycerol for high-value products.

(3) **Source of vegetable oil**

A variety of vegetable oil can be used as the source of biodiesel, such as rapeseed and canola, oil palm, soybean, sunflower, peanut, flax, safflower, castor seed, tung, cotton, jojoba, jatropha and avocado.

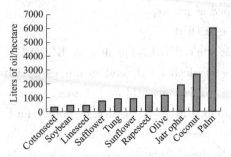

Figure 3-5　Productivity of oil per hectare land

Figure 3 – 5 presents the oil yield values as L/hectare for those crops. For other applications, some crops are grown for fiber and protein purposes. For example, soybean has been grown for protein with only 18% oil content.

Rapeseeds adapts very well in soil with poor fertility, however yielding 40% ~ 50% oil. Rapeseeds are the most important raw materials for biodiesel production in European countries. However, technological limitations still exist for sowing and harvesting in some Central and South American countries. This is because limited information is available for those countries in terms of fertilization, seed handing and storage. Another important factor is low price, in comparing with wheat and low productivity per hectare land. Rapeseeds also have high nutritional values compared with soybeans. Thus they are often used for protein supplements in cattle rations.

Canola represents "Canadian oil low acid". Canola is genetic modified rapeseed to reduce the content of erucic acid and glucosinolates in rapeseed oil, which have caused issues in animal and human consumption. Canola is also known as one of the best cooking oil with olive oil.

Soybeans have been largely grown in East Asian countries, as well as American countries such as U.S., Brazil and Argentina with vast variations in different environmental conditions. Obviously growing soybean will produce biodiesel as well as large quantities of valuable co-products, such as pellets and flour. Grain yields can range from 2,000 ~ 4,000 kg/hectare. The oil yield is approximately 18%.

Oil plan is a type of plant with a life span of approximately 25 years and typical height of 20 m. Two types of oil product can be produced from oil palm, palm oil proper from pulp and kernel oil from nut. Indonesia and Malaysia are the leading producers for palm oil. The demand for palm oil has been growing steadily in the past decades, mainly due to the increasing application for margarine and bakery products. Pure palm oil is actually in semi-solid form at room temperature. It is easily transportable and mixed with other vegetable oils to meet requirement of various applications. Sometimes it can be hydrogenated to yield saturated fatty acid esters.

Sunflower seeds are among most popular fruits especially in Asian countries. The seeds contain high values of nutritional components and good tastes. The oil yield is in the range of 48% ~ 52% with low content for linoleic acid, thus can be stored for a long period of time.

It is particularly noted that, linseed oil from flax seed contains high quantities of polyunsaturated fatty acids. Flax oil can produce as high as 40% ~ 68% of linolenic acid of the total production. Flax plant adapts the environment very well from wet to dry conditions, with low yields in seed and fibers.

The grain yield of safflower per hectare is low, although this plant can adapt the dry environment very well. The oil yield is 30%~40%, therefore it is economic to grow safflowers in arid areas. Two categories of valuable oil can be obtained from safflowers, mono-unsaturated fatty acids (mainly oleic acid) and other poly-unsaturated fatty acids (linoleic acid). The typical property for safflower derived oil is low cholesterol content. In addition to human consumption, safflower oils are often used in paints, coating, lacquer and soap.

3.3 Fuels from cellulosic biomass

Biomass conversion to fuels is known as the most longest history in human society. Chemical and biological transformation of cellulosic biomass is important to produce sustainable fuels. Thermal processing mainly consists of three types of pathways, pyrolysis, aqueous phase reforming and gasification.

Gasification is the oldest and most developed technique for fuel production. It is typically conducted at high temperatures (600~900℃) in the presence of controllable amounts of air or oxygen and steam. The product gas is mainly CO and H_2. Catalytic gasification can produce liquid fuels from syngas. Two types of strategies can be employed. One is above-mentioned fermentation or catalytic conversion of syngas to ethanol and higher alcohols. The second type is catalytic conversion of syngas to alcohols or alkanes.

Pyrolysis produces a mixture of bio-crude or bio-oil intermediates through moderately high temperature reaction in the absence of oxygen. The products, bio-oils contains hundreds of different compounds rich in oxygen species. The advantage for pyrolysis is that the resultant fuels often exhibit high energy density, which could be transported from remote sites to refining units. In particular, slow pyrolysis, fast pyrolysis and liquefaction are three major production methods. Slow pyrolysis is conducted at 450~500℃ with longer contact time under ambient pressure to generate bio-oil, gas and char. Fast pyrolysis is achieved by rapid heating rates at similar temperature ranges with 1~2 s contact time. Biomass feedstocks need to be dried prior to pyrolysis processes. Liquefaction can be carried out using water at elevated pressure (12~20MPa) in the absence of oxygen. The bio-oil produced in this process has relatively lower oxygen content compared with that derived from fast pyrolysis.

Aqueous phase reforming leads to the formation of hydrocarbons, oxygenates and H_2 from aqueous solution containing sugars and polyols. Reforming reactions occur in the presence of different solid catalysts such as Pt/Al_2O_3 and Pd/SiO_2. Gasoline range hydrocarbons and H_2 can be directly produced from this process. Further processing can also generate diesel and jet fuel range products.

3.3.1 Difference between fossil-and bio-derived fuels

Modern chemical industry has been established based on petroleum and coal conversion technologies. Production of transportation fuels from petroleum refining provides almost all essential fuel products including gasoline, diesel, jet fuel, kerosene and fuel oil. In this context, catalytic cracking, isomerization, hydrocracking, hydrodesulfurization have well developed technologies for energy industry. High energy intensity and complicated flow schemes are main flaws for conventional fuel processing technologies. Development of suitable catalysts and efficient reaction systems will be indispensable in facilitating future sustainable transformation technologies. In the field of biomass conversion to fuels, alternative technologies have been used, while similar principles should still be considered.

Figure 3-6 presents the key aspects for sustainability of chemical processing technologies. Clearly, safety and health, environmental impact and economics are critical supporting factors for sustainability. For example, biomass conversion to fuels requires renewable biomass as feedstocks. However, economics of biomass conversion processes should follow the green chemistry principles. It should be noted that, while oil refining in petrochemical industry mainly deal with C—C bond cleavage, and C—X (X: O, Cl, Br, *etc.*) bond formation, transformation of biomass to fuels is mainly focused on defunctionalization of bio-oxygenates. The fundamental difference in fossil fuel and biomass conversion has motivated researchers worldwide for advanced studies on catalyst design and reaction engineering.

Figure 3-6　Key aspects for sustainable chemical processing

Figure 3-7 shows the schematic illustration on typical operating temperatures and pressures involved in petroleum and biomass conversion. Detailed inspection on Figure 3-7 suggest that, petroleum refining processes are conducted on wide ranges of temperature and pressure. In particular, most petrochemical processes are actually carried out in vapor phase. Therefore, elevated temperatures are often needed to ensure gas-solid (catalyst) reaction occurring rapidly. Reforming of alkanes producing aromatics and derivatives is operated under relatively low pressure but very high temperature, to shift chemical equilibrium towards H_2 generation side. Liquefaction of coal and heavy oil needs both high temperature and pressure, to facilitate C—C bond cleavage reactions and formulate smaller hydrocarbon molecules.

Interestingly, biomass upgrading to fuels is carried out in different P-T zone in Figure 3-7. Owing to high oxygen content in cellulose and hemicellulose, for example, glucose, xylose, sorbitol and xylitol, hydrodeoxygenation or sometimes referred as hydrogenolysis (cleavage of

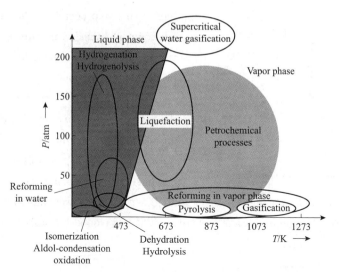

Figure 3-7 Operating pressure and temperature for biomass and petrochemical conversion processes

chemical bond) is required at relatively low temperature (<300℃) but relatively higher (H_2) pressure. Reforming of bio-oxygenates, particularly polyols, can be achievable in aqueous phase. The remarkably lower temperature for aqueous phase reforming is known as the most advanced technologies for H_2 production from biomass.

Based on the comparison shown above, the following points need to be carefully considered for biomass conversion to fuels. Is conversion of cellulosic biomass to fuel technologically viable? This is because C—O bond cleavage requires rational design of solid catalyst materials for good chemoselectivity. Another important question is that, is it economically sound? According to the reaction chemistry, removal of oxygen species results in a significant loss of weight for fuel products. Therefore, hydrodeoxygenation might be suffering from an uneconomic dilemma where large quantities of feedstock atoms are lost to form less valuable by-products.

3.3.2 Thermal conversion of biomass

As mentioned before, conversion of biomass to fuels has a long history in human society, much earlier than the establishment of petrochemical and coal industry. Direct thermal conversion of biomass/bio-wastes through heating is the most convenient approach. In general, direct thermal conversion includes torrefaction, pyrolysis, gasification, and combustion. It can also occur in the presence of a solvent, namely hydrothermal or solvothermal process.

Different from petroleum and coal conversion, biomass substrate often contain fuel in solid forms. Therefore, elevated temperature or oxidative species might be required to cleave large and viscous substrates into small portions. For example, combustion process involves drying, pyrolysis, char combustion and ash forming steps, as shown in Figure 3-8. Drying step actually physically remove the content of water species from solid fuel. Devolatilization release volatile compounds following drying stage. Continuing increasing temperature leads to significant formation of low molecular weight species and high molecular ones, through breakage of chemical bond in solid

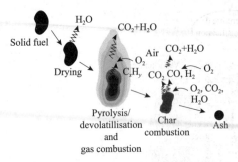

Figure 3-8 thermal combustions for solid fuels

fuels. This step occurs in the absence of oxidative environment. Char forms after devolatilization step, which can be further converted into CO, CO_2, H_2O via oxidation reactions.

Figure 3-9 presents temperature diagram for thermal conversion of solid bio-fuels. In particular, reaction network and possible products during torrefaction, pyrolysis, gasification and combustions. Increasing temperature up to 250 ~ 300℃ results in torrefaction/palletization to form bio-coal as intermediate product. Torrefaction process results in a reduction in mass of as high as 30% (wt) on a dry basis. Such loss also causes roughly 10% in energy loss.

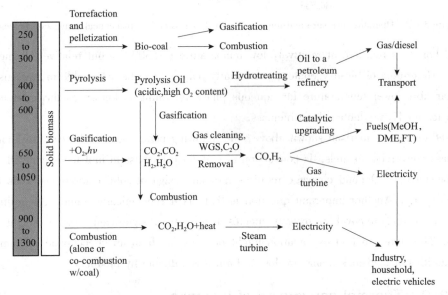

Figure 3-9 Temperature diagram and reaction network for various thermal conversion processes

Pyrolysis of solid fuels occurs when further elevating temperature to approximately 600℃. Pyrolysis shows several unique features for fuel processing. First of all, moderate temperature and rapid heating of biomass particles with short residence time could be beneficial for equipment design. The yield of bio-oil is approximately 75%, mainly including water-containing red-brownish oxygenated organic liquid. The char yield is about 15% with 10% of energy in biomass. It is important to mention that, oxygen removal is not significant during pyrolysis process. Thus, acidic products containing high oxygen content are formed. This type of product is often referred as bio-oil. Bio-oil can be further hydrotreated as important feeds for petroleum refining. Various types reactor have been applied for pyrolysis processes, including bubbling fluidized beds (Figure 3-10), transported bed, entrained flow reactors (Figure 3-11) and vacuum reactors. Typically, bio-oil consist of 54%~58% (wt) C, 5.5%~7.0% (wt) H, 35%~40% (wt) O. Heating value is about 16 ~ 19 MJ/kg. Bio-oil can be tested in diesel engines, turbines and boilers. The following aspects should be seriously considered. No char should be present. Nozzle, pressure values and other seal parts should be acid resistant and tolerate low pH values.

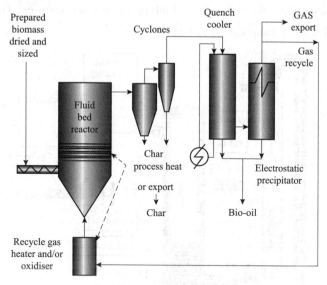

Figure 3 - 10 Pyrolysis processes using fluidized bed

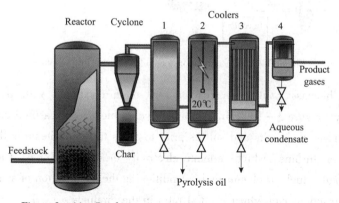

Figure 3 - 11 Pyrolysis processes using entrained flow reactor

In the meantime, enhancing reaction temperature to 1000℃ causes formation of CO_2 species, possibly through decarboxylation reactions of bio-oil, generating significant amounts of CO and H_2 products. Gasification occurs at such elevated temperature and generate large amounts of light gases in the presence of O_2 or air (Figure 3 - 12). In general, solid fuels and relatively heavy oil undergo cracking to form CO_2, CO, H_2, H_2O, CH_4, H_2S and tars ($C_xH_yO_z$). Relatively higher molecular weight molecules are undesirable products from gasification because they may condense and cause fouling in downstream processes. Gasification reactor technologies include fluidized beds (bubbling and circulating), moving beds (cocurrent and countercurrent), grate (both moving and stationary), and entrained flow. Fluidized bed reactors are usually operated under melting temperature of ash to avoid bed agglomeration. Moving beds can be operated around melting temperature while entrained flow reactors are operated above melting temperatures. The syngas produced from a direct gasification process represents approximately 70% of energy in the starting biomass materials. Air-blown gasification reactors can produce syngas mixtures with 6 ~ 10 MJ/Nm^3 in energy density. Since N_2 in the air is diluting the product mixtures, O_2-blown gasification syngas can yield a higher heating values of > 10 MJ/Nm^3.

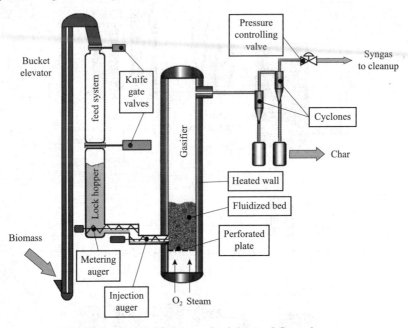

Figure 3 – 12　Gasification technologies and flow schemes

In practice, biomass gasification reactors are often combined with power generation to generate electricity (Figure 3 – 13). In combined cycle, biomass is gasified and syngas mixtures are cleaned to remove alkali salts and sulfurs prior to a gas turbine. Steam is then generated with hot flue gas from gas turbine, which generate electricity in steam turbines. Catalytic gasification can be used to produce fuels or chemicals by conditioning the composition of syngas. Incorporation of water gas shift reactions can adjust H_2/CO ratio in the products.

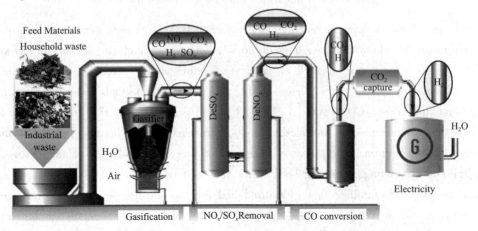

Figure 3 – 13　Gasification of household wastes and generation of electricity

In combustion operation, sufficient air or O_2 is introduced to ensure complete oxidation of fuels. Complete oxidation of biomass can be combined with electricity generation plants for both power and heat production. Dedicated furnace or co-firing boilers can be operated for fluidized bed reactors (Figure 3-14). Operation temperatures for boilers are often limited by melting points of ashes. Typical electricity generation efficiency is around 40% of thermal energy in burned fuel.

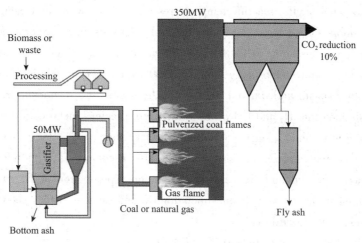

Figure 3-14   Combustion of biomass and waste

Direct thermal conversion is one of the most traditional ways of using bio-energy. The main purpose for torrefaction is to improve energy density and handing solid biomass for combustion and gasification units. Pyrolysis processes generate liquid fuels with high oxygen content. Temperatures for pyrolysis should be high enough to break down solid fuels to high molecular weight molecules (tars). Gasification processes can produce syngas and power for operating plants. Combustion units are designed for complete oxidation of biomass for power generation.

3.3.3   Key platform intermediates from biomass

Identifying key compounds for conversion of cellulosic biomass to fuels and chemicals is extremely important. U. S. Department of Energy has published reports on 12 key platform compounds for biomass conversion. International Energy Association reported Task 42 for biorefinery classification system.

(1) H_2 as an energy carrier

Researchers from U. S. have conducted comprehensive screening on possible platform compounds for catalytic synthesis of bio-fuels. This route is different from direct thermal conversion. Two possible conversion routes have been studied for catalytic conversion of bio platform compounds to fuels and additives. This part of work is particularly investigated for cellulose and hemicellulose. Since conversion of glucose to bioethanol has been largely implemented as first generation technologies, direct conversion of cellulose to ethanol has been regarded as one of the most promising technologies for future biofuels. Since early 21st century, direct conversion of sugars and polyols to H_2 and hydrocarbons has gained worldwide attention.

Aqueous phase reforming of sugars or polyols containing solution in the presence of solid catalysts display remarkable technological advantages over conventional ways of H_2 production. A typical biomass feedstock, glycerol, obtained from biodiesel (as already discussed) can undergo reforming reaction to produce H_2 and CO. The importance of this work lies in the reaction condition and new chemistry involved in glycerol conversion. Reforming processes in petrochemical industries requires much higher reaction temperatures (>500°C) with short catalyst life owing to the poisoning of noble Pt catalysts. In sharp contrast, aqueous phase reforming occurs at only 200~250°C in liquid medium. This finding is considered as a major breakthrough in H_2 production technology. More importantly, this is a 100% green H_2 production process.

Similarly, other sugars and derived polyols, such as glucose, mannose, fructose, sorbitol, can also be transformed into H_2 and CO in the presence of precious metal catalysts. Different from glycerol, reforming of C_5 and C_6 sugar polyols tends to formulate oxygenates rather than H_2. But it is necessary to note that, water gas shift reaction also accompanies with reforming reaction, especially with noble Pt, Pd and Ru catalysts. Therefore, production of H_2 can be maximized using noble metal catalysts. Non-noble ones including Ni and Co also display good performances in conversion and selectivity towards H_2.

(2) Transformation of oxygenates to alkanes

Transformation of sugars and polyols (carbohydrates) into hydrocarbons requires removal of oxygen species. This process demands addition of molecular H_2 to achieve hydrogenation or hydrodeoxygenation reaction. The use of H_2 should be minimized to avoid unfavorable increase in operation cost. Ideally, H_2 used in hydrodeoxygenation processes should be produced from renewable resources. Electrolysis of water and reforming, water gas shift reaction has not provided enough industrial H_2 for hydrodeoxygenation of bio-oxygenates. At present, industrial H_2 is still largely obtained from petroleum and coal.

As we know, hydrogenation of glucose, a monomer from cellulose polymer, yield sorbitol as the main product (selectivity $>99\%$). Sorbitol is much more thermally stable compared with glucose, which can undergo further transformations. Sorbitol is a C_6 linear polyol with one-OH group on each carbon. Dumesic and Huber are among first research teams to find that sorbitol can undergo selective hydrodeoxygenation to yield hexane as a major product, in the presence of Pt/SiO_2-Al_2O_3 catalysts. How to selectively facilitate C—O bond cleavage while restraining C—C bond breakage is important for hydrodeoxygenation of sorbitol. C—C bond cleavage reactions include retro-aldolization, decarbonylation on metal sites. Dehydration reaction is very important in this process, because sorbitol can form cyclic ethers, which can then be hydrogenated to form deoxygenated intermediates. Isosorbide, is the major product from sorbitol ring-closing dehydration and known as the precursor for hexane. Isosorbide can undergo ring-opening hydrogenation to form 1,2,6-hexanetriol as another important intermediate. Hexanol and hexane are both found to be the final products for sorbitol transformation over Pt/SiO_2-Al_2O_3 catalysts, at 240 + °C and 8MPa H_2. In additional to C_6 products, a variety of C_2—C_3 hydrocarbons and oxygenates are also formed simultaneously. The overall yield for hydrocarbons and hexanes are approximately 56% and 30%, respectively, suggesting additional research work is required to further improve the chemoselectivity for alkanes.

According to the stoichiometry, six equivalents of H_2 are required to convert one mole of sorbitol to hexane. Therefore, achieving good selectivity for C—O bond activation is a challenge for this process. If one combines reforming and water gas shift reactions with hydrodeoxygenation of sorbitol, the operating conditions could potentially be milder. One mole of sorbitol could produce 13 moles of H_2 through reforming and water gas shift reactions. Therefore, to produce one mole of hexane as target product, the theoretical stoichiometry is 19/13, roughly consuming 1.5 mole sorbitol.

$$C_6O_6H_{14} + 6H_2 \longrightarrow C_6H_{14} + 6H_2O$$
$$C_6O_6H_{14} + 6H_2O \longrightarrow 6CO_2 + 13H_2$$
$$\text{Sum } \frac{19}{13}C_6O_6H_{14} \longrightarrow C_6H_{14} + \frac{36}{13}CO_2 + \frac{42}{13}H_2O$$

But it is important to note that, the combustion, fluid and flash properties for hexane are intrinsically unfavorable for blending with existing gasoline pool. Despite promising features for hydrodeoxygenation applications, production of renewable hexane from sorbitol might not be commercially viable for fuel production.

(3) Catalytic conversion routes of cellulose-derived compounds for biofuels

Since linearly structured hexane product has many disadvantages in combustion, researchers were seeking alternative conversion pathways to biofuels starting from sugars. Huber and colleagues have developed a different conversion technologies involving 5-hydroxymethylfufural (HMF) as key intermediate compound. HMF can be derived from glucose through isomerization and dehydration reactions. In particular, isomerization of glucose yield fructose while dehydration reaction further leads to the formation of HMF in aqueous/organic medium. Similarly, furfural can also be obtained from xylose conversion. HMF derived biofuels have been widely studied in literature. Hydrodeoxygenation of HMF and derivatives can produce gasoline compounds or additives. Sequential aldolization of HMF with acetone followed by hydrodeoxygenation to obtain $C_8 \sim C_{15}$ diesel range products (Figure 3-15). The conversion processes include multiple steps starting from cellulose or hemicellulose materials. Hydrolysis of those bio-polymers yield glucose and xylose as main products. Isomerization and dehydration facilitated by acidic catalysts further convert sugars into furfurals. Low-temperature aldolization and hydrogenation or high-temperature hydrogenolysis/hydrodeoxygenation eventually give desired ranges of products.

Aldolization reaction can be catalyzed by homogeneous NaOH or heterogeneous basic metal oxide such as MgO, CaO, BaO and Mg-Al oxides. Low-temperature hydrogenation is conducted over Ru/C (5% (wt)) catalysts at 120℃ and ~4MPa H_2 pressure. This step could saturate C=C and C=O bonds in the product molecules. High-temperature deoxygenation can be achieved using Pt/SiO$_2$-Al$_2$O$_3$ catalysts at elevated temperatures and pressures (260℃ and >6MPa).

Through aldolization processes, approximately 90% yield of jet fuels can be achieved in terms of hydrocarbons. For example, about 0.45 kg jet fuels can be produced per kg of xylose.

Hydroalkylation is another important reaction to add carbon chains to existing oxygenates.

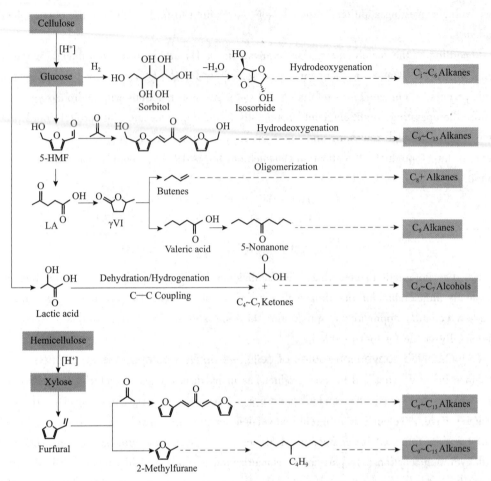

Figure 3-15 Roadmap for conversion of cellulose to hydrocarbons

Corma and colleagues studied hydroalkylation of 2-methylfurfural with butanol catalyzed by soluble *para*-toluolsulfonic acid or solid Amberlyst-15 catalysts. Actually 1,1-bisylvylbutane can be easily obtained and further hydrogenated into hydrocarbons with remarkable carbon selectivity (95%) using Pt based solid catalysts. Ring opening reactions can also occur over 2-methylfurfural to generate 4-oxopentanal in the presence of acid catalysts. This intermediate then reacts with other molecules to generate trimers, which could be further hydrodeoxygenated into hydrocarbons. The as-obtained products in organic phase have remarkable cetane number of 70.

Levulinic acid (LA) is another important intermediate, as it can be generated as a by-product owing to unfavorable decomposition of HMF in aqueous medium (Figure 3-16). Utilization of LA has been a hot topic, as it can be used as an important bio-solvent in numerous applications. In addition, formic acid is also released with LA during production, which could be used as H donor in chemical industry. Decomposition of formic acid to H_2 and CO_2 for transfer hydrogenation reactions is regarded as a promising technology in green chemical engineering.

Hydrogenation of LA gives γ-valerolactone (GVL) as the main product. Conversion of GLV into C_8+ hydrocarbons can be realized in the presence of externally added H_2. In this route, GVL

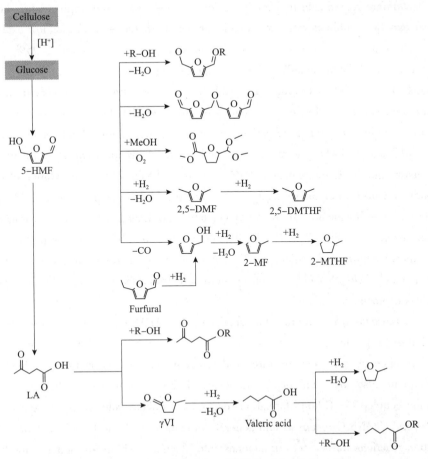

Figure 3-16 Valorization of HMF and levulinic acid

can be transformed into pentenoic acid and undergoes decarboxylation to form butane for sequential oligomerization reactions to eventually generate C_8 hydrocarbons. To achieve those multiple reaction steps, aqueous phase containing GVL can be isomerized into pentenoic acid in the presence of SiO_2-Al_2O_3 solid acid catalyst. Isomerization and decarboxylation reactions can be realized over such solid acid catalyst while coke formation and associated catalyst deactivation need to be resolved. Water component needs to be removed to mitigate the inhibitory effect on catalyst surface for oligomerization reactions. Isomerization and decarboxylation reactions can be combined as one single stage at 380℃ and 4MPa.

GVL can be further transformed into valeric acid in the presence of metal catalysts. This reaction can be achievable over bifunctional Pd/Nb_2O_5 catalyst under moderate temperatures and pressures. This route produces valeric acid as an important precursor for biofuels. Transformation of GVL to valeric acid can be catalyzed by bifunctional catalytic materials with both hydrogenation and acidic sites. Pt/ZSM-5 catalysts have been designed for this process with a total valeric acid yield of 90% at 250℃ and 1MPa H_2. Balancing the activity for hydrogenation and acidic sites is important for selective formation of valeric acid. Excess amounts of hydrogenation sites might lead to the formation of 2-methyl tetrahydrofuran.

Ethyl esters are remarkable fuel additives for gasoline and diesel applications. Formation of valeric acid can be combined with esterification reaction in the presence of ion exchange resin catalysts. Valeric acid as formed can react with alcohols to form esters with >95% yields. The alkyl valerates have been found to show superior performances as gasoline and diesel. Approximately 15% (v) ethyl valerate can be blended with gasoline for combustion engines.

Valeric acid can be further converted to 5-nonane and CO_2 through ketonization reactions. Generation of valeric acid and ketonization reactions can be combined into a one-stage reactor system at 350℃ and 3.5MPa pressure. An organic layer of products can spontaneously separate from aqueous medium with approximately 60% yield of C_9 ketones. Decarboxylation of valeric acid is also conducted over CeZr oxide catalysts at relatively higher temperature (~430℃) with a yield of 84%. 5-Nonanone can be hydrogenated into linear nonane via hydrogenation and dehydration reactions, which has a good cetane number and lubricitive properties, making it suitable as diesel additive. Additionally, 5-nonanone can be hydrogenated over USY catalysts, yielding 5-nonanol. This alcohol can then undergo dehydration/isomerization to give hydrocarbons for use as gas components.

The above-discussed routes are established based on HMF chemistry. Lactic acid is another important industrial product, which is usually obtained through fermentation of starch and sugars. Lactic acid is often used for manufacture of degradable plastics, cosmetics and food additives. Lactic acid can also be catalytically transformed into hydrocarbons through dehydration/hydrogenation coupled C—C bond formation reactions. Aqueous solution containing lactic acid is processed overall Pt/Nb_2O_5 catalysts and generate propanoic acid and $C_4 \sim C_7$ ketones. Sequential hydrogenation reactions further convert ketones into alcohols, which are suitable for liquid fuels with high energy density. Experimental studies have revealed that, Nb_2O_5 supports provide active sites for dehydration and C—C coupling while Pt sites are active for hydrogenation reaction.

Similarly hemicellulose can be transformed into xylose and furfural as important precursors. Furfural undergoes aldolization to form $C_8 \sim C_{13}$ alkanes for transportation fuels, while 2-methylfurane can undergo C~C coupling to form $C_9 \sim C_{15}$ hydrocarbons.

Additional discussion on HMF conversion will be carried out in this section. In addition to aldolization, hydrodeoxygenation of HMF can lead to formation of 2,5-dimethylfuran (2,5-DMF) or 2,5-dimethyltetrahydrofuran (2,5-DMTHF). This route has been investigated by Dumesic's group. The oxygen content in HMF can be further reduced in this process. DMF and DMTHF are excellent fuel additives. In particular, DMF exhibits a remarkable energy density of 31.8 MJ/L and boiling point of 90~92℃, making it suitable for fuel applications. It is important to note that, the energy density for DMF is approximately 40% higher than ethanol.

Production of DMF can be realized in a continuous reactor using fructose as the starting material. Dehydration of fructose to HMF is achieved in a biphasic system, subsequently, HMF being extracted by organic phase and converted to DMF through hydrogenolysis over CuRu catalysts. When butanol is used as organic solvent, the selectivity of HMF is 82% at 85% conversion. Hydrogenolysis reactions to DMF leads to a 76%~79% yield over CuRu/C catalysts.

3.4 Lignin for fuels and chemicals

3.4.1 Structures of lignin

A typical wood material of heterogeneous, hygroscopic and cellular properties consists of cells and cell walls (Figure 3-17). Cell walls mainly consist of cellulose (40%~60%), hemicellulose (~34%) and lignin (~30%). Situated among cellulose and hemicellulose in the plant secondary cell wall, lignin acts as structural glue that gives terrestrial biomass structural rigidity. Lignin supports structure coherence through defending against chemical and biological attack.

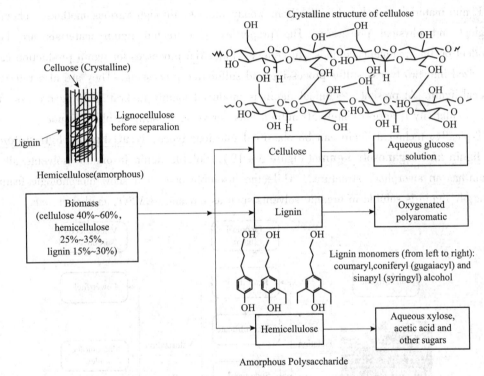

Figure 3-17 Structures of lignocellulosic biomass

Indeed, lignin is a heterogeneous and amorphous biopolymer that is composed of a series of linkages formed by radical coupling reactions (Figure 3-18).

Characterization of lignin structures is critical in efficient utilization for chemicals and fuels. Various analytic methods, such as FTIR, NMR have been widely employed in revealing the structures of lignin. The formation of lignin is the result of radical coupling/polymerization of three main **monolignols**, **sinapyl**, **coniferyl** and ***p*-coumaryl** alcohols. Vast majority of lignin is composed of the three units. Lignin is amorphous macromolecular structures composing of the three units, linked mainly by ether linkages and condensed linkages. About 40%~60% of overall intermolecular linkages in lignin are ether bonds and the β-O-4 bond is a predominant ether linkage.

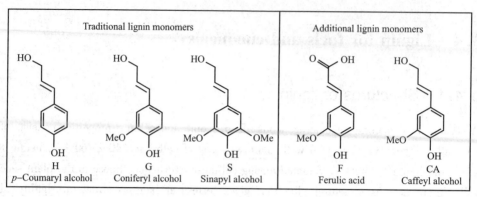

Figure 3 – 18 Lignin monomers

Lignin materials can be obtained from woody biomass through various methods, chemical, biological and physical processes. The properties of extracted lignin materials are largely dependent on the treatment methods. In general, industrial processes for lignin production can be categorized into two types: sulfur processing and sulfur-free processing. They are also referred as technical lignin. At present, technical lignin is produced mainly in kraft pulping process. Kraft lignin productivity is approximately 50 million tons per year, existing in black liquor.

Typically, technical lignin can be classified into four types, **Kraft lignin**, **lignosulfonate**, **soda lignin and organosolv lignin** (Figure 3 – 19). Soluble lignin in organic solvents/alkaline medium has an amorphous structures, while the insoluble one often show morphologic features. Organosolv lignin is soluble in organic solvents such as dioxane, DMSO, methanol, *etc*.

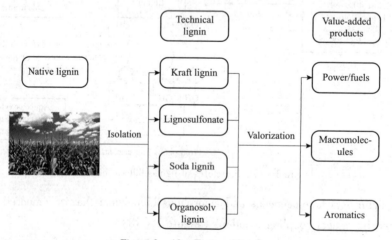

Figure 3 – 19 Types of lignin

(1) Kraft lignin

The choice of lignin production (separation) has a great impact on bio-refinery. Generally speaking, Kraft lignin is relatively more attractive, as virtually all lignin process streams worldwide are based on Kraft processes. Kraft lignin is produced at high pH in the presence of substantial amounts of aqueous sulfides at temperatures of 150~180℃.

Black liquor obtained from Kraft processes is used to supply power for steam and recovery units. Black liquor is normally concentrated up to 40%~50% solids and then burned for heating values (about 12,000 ~ 13,000 BTU/dry lb). One way to efficiently utilizing Kraft lignin is to implement large-scale gasification units. The gas turbines of lignin gasifiers are twice as efficient as normal steam turbines.

Kraft lignin can be recovered from black liquor by lowering pH values. Substantial amounts of Kraft lignin are precipitated under low pH environment. Considering the high concentration of sulfides in aqueous medium, sulfur content for Kraft lignin is only 1%~2%.

The main portion of Kraft lignin is isolated chemically as sulfonated products in two processes (Figure 3 - 20). One process involves reactions at 100 ℃ to introduce sulfonate functionality into aliphatic side chains. Another step introduces sulfur species by sulfonation occurring on aromatic rings at 150 ~ 200 ℃. Alternatively, acid induced hydroxymethylation of lignin's aromatic rings with formaldehyde can be achievable.

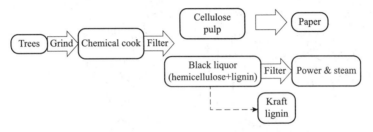

Figure 3 - 20　Production of kraft lignin

(2) Lignosulfonate

Worldwide production of lignosulfonate is approximately 1 million t/a. Sulfite pulping is carried out between pH 2 ~ 12, depending on the cationic composition of liquor. Most sulfite pulping uses calcium or magnesium as the counterion. Pulping as high pH is conducted with sodium or ammonium conterions. Lignosulfonates are soluble within wide ranges of pH values. Therefore then cannot be separated by adjusting pH values. Thus lignosulfonates are often recovered from waste pulping liquor concentrates after stripping and recovery of sulfur (Figure 3 - 21). Precipitation of lignosulfonate calcium salt is the simplest method. Lignosulfonates are typically higher molecular weighted compared with Kraft lignin, of 1,000 ~ 140,000 in average.

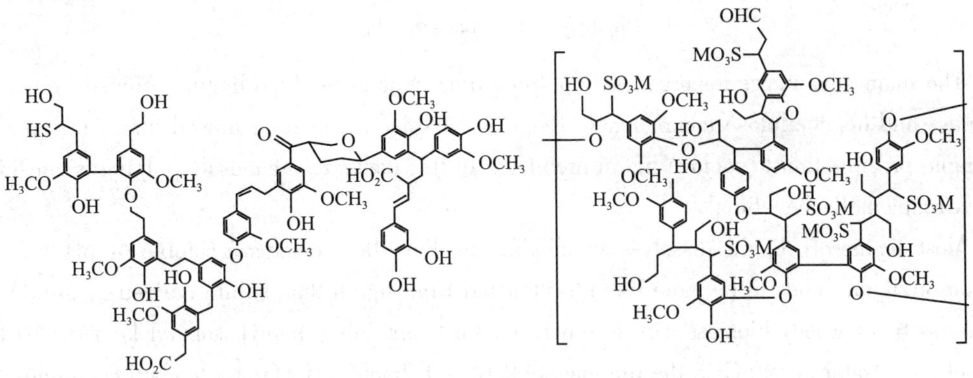

Figure 3 - 21　Kraft lignin and lignosulfonate

Despite of high functionality, lignosulfonates only account for 2% of total pulp production. Lignosulfonates are also contaminated by cations during pulp production and recovery. Lignosulfonates are only 70%~75% in lignin structures, with remaining species primarily as carbohydrates and inorganic materials.

(3) Organosolv lignin

Organosolv pulping is a general term for the separation of wood components through treatment with organic solvents. For example, extracting lignin from wood can be realized involving heating in aqueous dioxane at elevated temperature. The ether linkage of β-O-4 bond for inter-unit can be largely retained under such condition (Figure 3-22). Various organic solvents and combinations can be applied for organosolv pulping. Acids and alkalis can enhance pulping rates. The most well-known Allcel process uses ethanol-water solvent system.

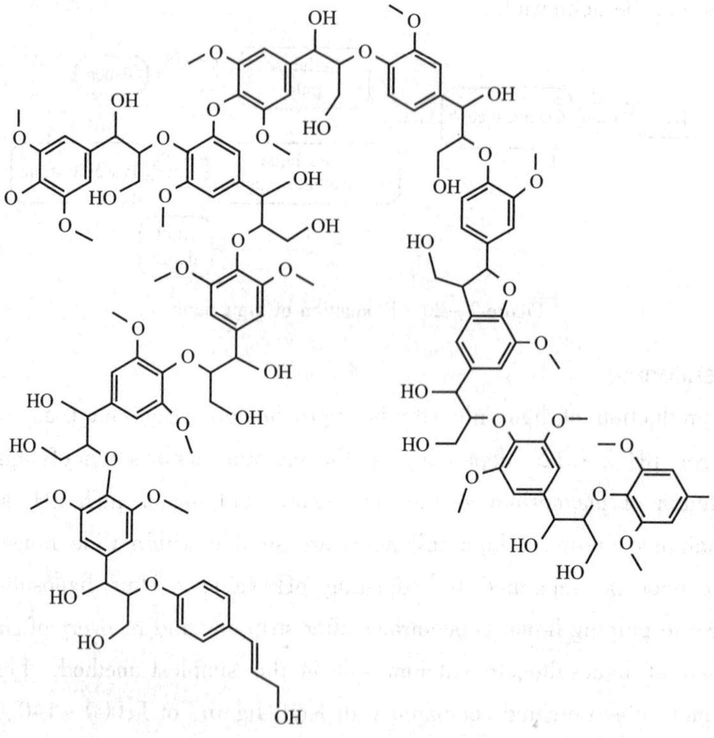

Figure 3-22 Organosolv lignin

The main advantages for organosolv pulping process is to produce lignin easily separated from cellulose and hemicellulose components. Recovery of solvent can be achieved through evaporation or simple precipitation to remove lignin products. In this context, organosolv pulping is considered as environmentally beneficial.

Most organosolv lignin dissolves in alkaline medium but remains insoluble at pH = 2~7. Organosolv lignin tends to be more like unsulfonated Kraft lignin than sulfite-derived lignosulfonate. A process treats woody biomass with a ternary mixture containing methyl isobutyl ketone (MIBK), ethanol, and water at 140℃ in the presence of 0.05~0.2mol/L H_2SO_4 as standard conditions. The solvent mixture selectively dissolves the lignin and hemicellulose components leaving the cellulose undissolved. Recovery of organic solvents is the economic determining factor for organosolv pulping processes.

(4) Other lignins

Pyrolysis lignin, steam explosion lignin, acid lignin and alkaline oxidative lignin are among other important types of lignin materials. Using carefully controlled fast pyrolysis technologies, bio-oil can be produced *via* short contacting time in reactor. A total yield of 75% (*wt*) can be obtained on the basis of dry woody biomass. Steam explosion lignin is produced using high pressure steam (3MPa) at short residence time. About 90% of lignin materials can be recovered after washing with alkaline species.

Acid treatment of biomass has been long developed for bio-refinery. In this process, cellulose and hemicellulose are separated from lignin and hydrolyzed into fermentable sugars. The main disadvantage is degradation of sugars into furfurals and carboxylic acids. Alkaline oxidative lignin materials are obtained by introducing O_2 or H_2O_2 in alkaline medium.

3.4.2 Depolymerization of lignin

Depolymerization processes involve various methods, such as biological catalysis using enzyme, and thermal conversion using heat or thermochemical approaches. Thermochemical approaches often include hydrolytic depolymerization using water and alkaline species/catalysts, reductive depolymerization using H_2, oxidative depolymerization.

(1) Hydrolytic depolymerization

Hydrolytic depolymerization process is performed in sub or supercritical water medium in order to achieve cleavage of ether bond in lignin materials. Acidic or alkaline catalysts are often required to facilitate such reactions typically at 280 +℃ and 20 + MPa pressure. A variety of monomers and subunits species are generated, such as phenols, catechol, guaiacol, and other methoxy phenols. Further degradation of methoxy groups also occurs under such condition, while the benzene ring is still stable.

But solid residues and char can be formed owing to repolymerization and condensation reactions taking place under the harsh reaction temperature and prolonged reaction time. Experimental studies demonstrate that addition of alkaline catalysts such as NaOH, K_2CO_3 and Ca$(OH)_2$ can reduce the formation of chars and facilitate depolymerization reactions. Mixing water with organic solvent can also prevent the formation of char and facilitate depolymerization. Water-ethanol, water-methanol, water-butanol, water-phenol and water-acetone are popular combination of solvents.

Model compounds have been used to elucidate the possible mechanism for lignin degradation. For example, hydrothermal treatment of guaiacol to catechol, phenol and o-cresol was conducted with maximum yield of catechol being 41%. Catechol can be further degraded into phenol in the presence of more water content at 373℃. It is also found that, the activation for catechol conversion is slightly lower (39 kJ/mol) under supercritical water compared with subcritical condition (50.7 kJ/mol).

The addition of alkaline catalysts can reduce depolymerization temperature by almost 100 ℃. For example, Lignin can be hydrolytically depolymerized in water-ethanol system in the presence of NaOH at 220~300 ℃. The molecular weight of alkaline lignin decreases significantly from M_w: 60,000 g/mol (M_n: 10,000 g/mol) to M_w: 1000 g/mol (M_n: 450 g/mol) at around 260 ℃.

Lignin materials can also be mixed with solid acid catalysts and converted in water-butanol system (Figure 3-23). Organosolv lignin and Kraft lignin was subjected to hydrolytic depolymerization using water/butanol mixture (4/1 molar ratio) over SiO_2-Al_2O_3 catalysts at 260~350 ℃. Depolymerization of lignin in aqueous phase over the Lewis acid sites of SiO_2-Al_2O_3 catalysts forms lower molecular weight liquid products such as monomers, dimers, and oligomers, while butanol phase can extract those intermediates and further fractionate to smaller molecule.

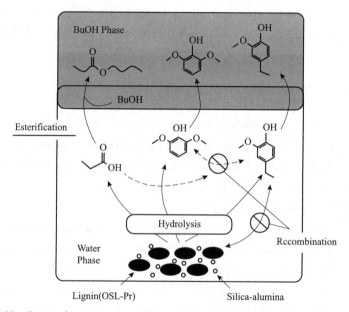

Figure 3-23 Proposed reaction routes for water-butanol system in depolymerization of lignin

(2) Reductive depolymerization

This type of process can be realized using noble metal catalysts in the presence of molecular H_2 at temperature above 300 ℃. The liquid products are mainly composed of volatile hydrocarbons, phenol and methyl-, ethyl-, and propyl substituted phenol, catechols, and guaiacols.

Reductive depolymerization is often combined with fractionation method to obtain targeted phenols. For example, hydrogenation reaction in combination with solvolysis in methanol solvent can lead to deoxygenated monomers. Various other supported metal catalysts including Ni/ZSM-5, RuCu/HY, NiCu/Beta and CuMo/ZSM-5 catalysts can convert organosolv lignin, Kraft lignin, corn stover lignin into aromatics by hydrogenolysis and hydrodeoxygenation reactions (Figure 3-24).

Degradation in the presence of noble metal catalysts follows alternative reaction pathways. Catalytic C—C cleavage could occur over Pd catalysts (Figure 3-25), while dehydrogenation is strongly believe to initiate the catalytic turnovers and provide active H_2 and intermediate for O—H cleavage or dehydration reactions.

Chapter 3 Biomass Conversion to Fuels and Chemicals

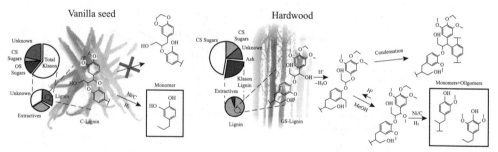

Figure 3-24 Schematic presentation of reductive-extractive-depolymerization of lignin materials

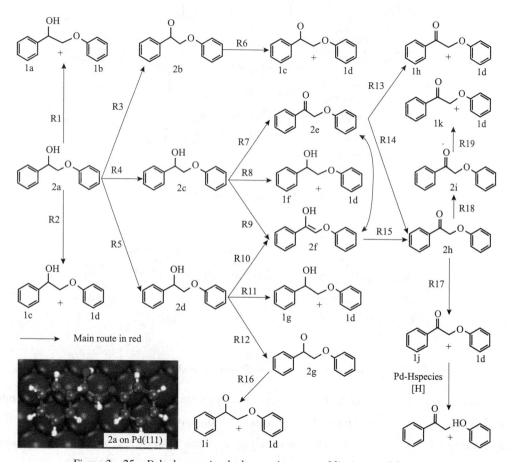

Figure 3-25 Dehydrogenation-hydrogenation route of lignin over Pd catalyst

Hydrogenation conversion of lignosulfonate occurs following activation of C—O—C bond to form phenols and methylene structures. Cleavage of hydroxyl groups at side chains produces C_1—C_3 alkyl substituted guaiacols and major products (Figure 3-26). C—S bond in lignosulfonate can also be reductively cleaved on metallic sites to form H_2S as final product.

Pd catalysts can also achieve transfer hydrogenolysis in MTBE water system (Figure 3-27), using formic acid or amines as H donors. The reaction mechanism seems to be similar as in the presence of externally added H_2. The introduction of stoichiometric H donors can easily facilitate transfer hydrogenolysis.

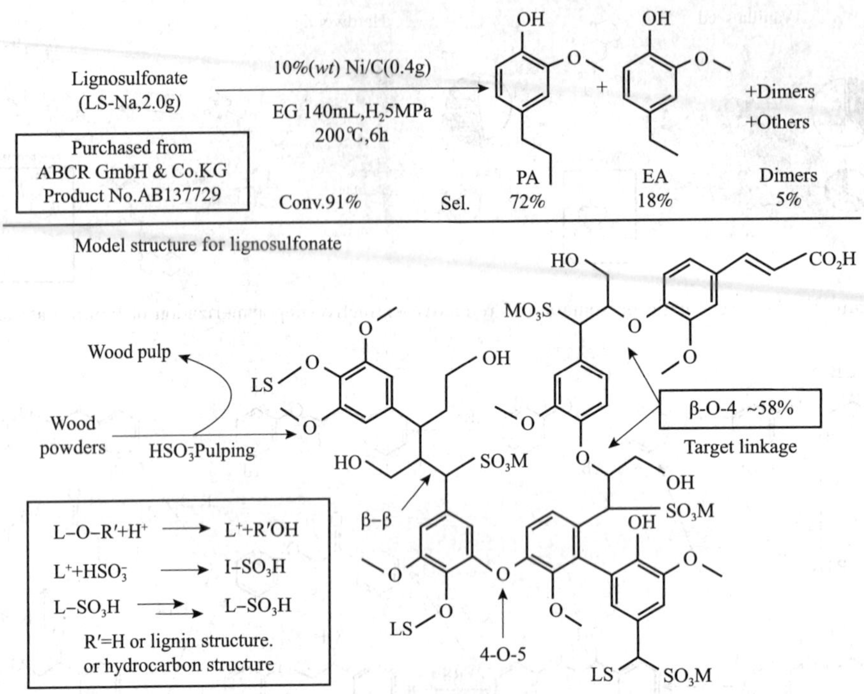

Figure 3-26　Conversion of lignosulfonate in reductive depolymerization

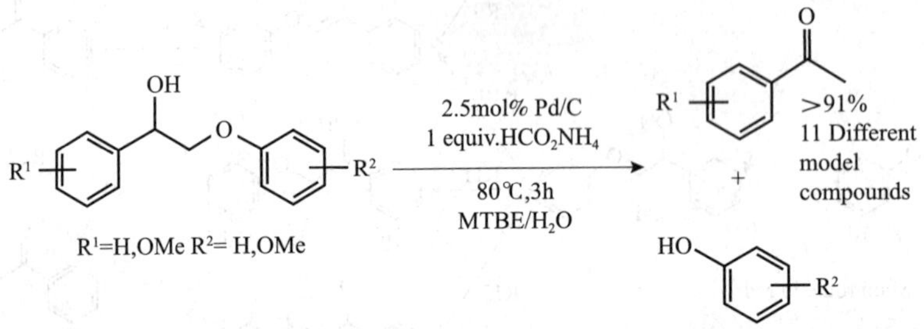

Figure 3-27　Pd catalyzed C—O bond cleavage of β-O-4 lignin model systems

Solvent also play an important role in reductive depolymerization, as alcoholic solvents often act as H donors for transfer hydrogenolysis processes. Protolignin alcoholysis study over the Ni/C catalyst was carried for fragmentation-hydrogenolysis of lignin. Methanol, ethanol and ethylene glycol can work as H donors and nucleophilic reagent for cleaving ether linkages to achieve the preliminary fragmentation of birch wood into smaller soluble species consisting of several aromatic rings. Ni/C catalysts can further hydrogenate the aromatic intermediates to phenols, with selectivity being 90% at 200℃. Formic acid can be used as effective reducing agents (Figure 3-28). While direct depolymerization of lignin only leads to <10% yet of low molecular aromatics, conversion of aspen lignin in the presence of formic acid achieves 52% efficiency at 110℃.

Solvent also affects the solubility of lignin and its intermediates. Experimental studies have revealed that, protonic polar solvents (e.g., alcohols) exhibit a higher dissolution capacity than do nonpolar solvents and the solvents with more -OH groups are better for lignin dissolution because

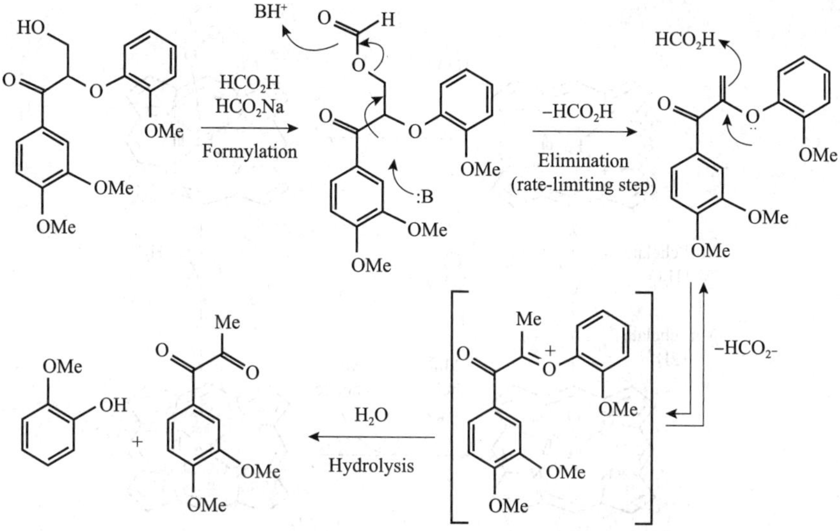

Figure 3-28   Cleavage of C—O bond using formic acid as H-donor

the -OH groups can weaken or replace H bonds to promote lignin releasing from lignocellulose and etherification with alcohol can further increase the dissolution capacity. Different from methanol and ethanol, 1,4-dioxane does not have -OH groups. Thus 1,4-dioxane dissolved lignin fraction is beyond 4000 in molecular weight, which is much higher compared with alcohol dissolved lignin fragment.

(3) Oxidative depolymerization

Naturally existing enzymes can degrade lignin materials under mild conditions, such as lignin peroxidase and manganese-dependent peroxidase with in situ generation of H_2O_2 (Figure 3-29). Both of the enzymatic catalysts work through a Fe^{3+} metalloporphyrin prosthetic group for lignin depolymerization. During oxidation reaction, Fe^{3+} metalloporphyrin undergoes a two-electron oxidation with H_2O_2 to yield oxo-Fe^{4+} species with π-cation radical species. The active intermediate then conducts two one-electron transfer reaction with lignin substrates to complete the catalytic cycle.

Researchers have also developed iron porphyrin catalysts for the oxidation of β-1 model compounds, similar to biological analogues, to cleave the C_α—C_β bond, leading to monomeric products. Iron tetraphenylporphyrin chloride and *tert*-butylhydroperoxide (TBHB) catalyst system has been applied to model compounds, also displaying good performances in C—C bond cleavage in producing monomeric aldehydes or acids. This is different from conventional oxidation methods where main reactions occur for benzylic position to form ketone products. Mn-porphyrin/H_2O_2 system has also been successfully applied for lignin oxidative depolymerization.

Cobalt Schiff base catalysts such as cobalt salen complexes are remarkable ones for oxidation reactions. Co(salen) catalysts can facilitate oxidation reactions by forming Co-superoxo complexes in the presence of molecular O_2 (Figure 3-30). In addition, the coordination of Co species also affect the oxidation activity. This is because that formation of aldehydic C=O group will undergo coordination prior to form active super oxo-Co(salen) species. Poor coordination of aldehydes is

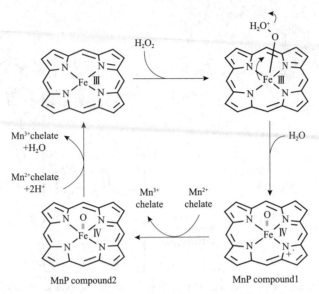

Figure 3-29 Catalytic cycle involving Mn and Fe catalysts

caused by stabilization effect by aromatic ring. Subsequently, the formation of a second oxo-Co (salen) species tend to attack the phenol like intermediate and eventually form benzoquinone. Finally, the cobalt species either breaks down to regenerate the starting catalyst (L_nCoOOH in Figure 3-30) or can act as catalyst in its own right (L_nCoOH).

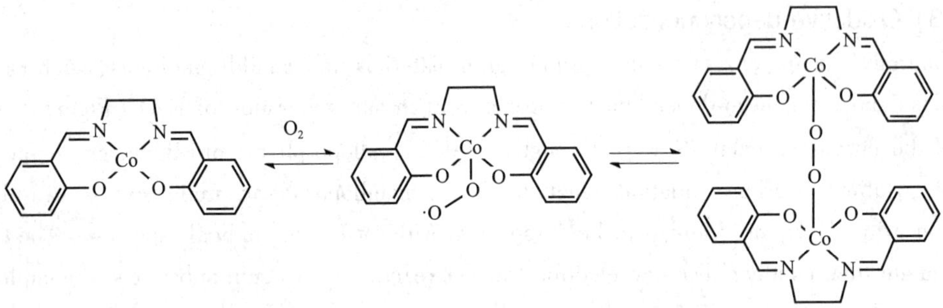

Figure 3-30 Formation of Co-superoxo complexes

Stepwise oxidative depolymerization can be classified as benzylic oxidation and secondary depolymerization. TEMPO (2,2,6,6-tetramethylpiperidine-N oxyl) and DDQ (2,3-dichloro-5,6-dicyano-1,4-benzoquinone) have been found to be ideal candidates for breaking C_4—C_β bond, which is ideal for oxidation of benzylic alcohols (Figure 3-31).

Cleavage of C_α—C_β bond is key for oxidation of phenolic and non-phenolic model compounds. Experimental studies on CuOTf/TEMPO/2,6-lutidine/O_2 system have found that this reaction is dominant even for non-phenolic models (Figure 3-32). Two reaction pathways have been proposed and validated. The reaction is undergoing either through primary alcohol oxidation followed by retro-aldolization reaction or one-electron oxidation to break C—C bond. In either of the reaction pathway, the main products are aldehyde and phenol rather than benzylic ketones. Interestingly, when phenolic model compounds were used, the dominant product was benzylic

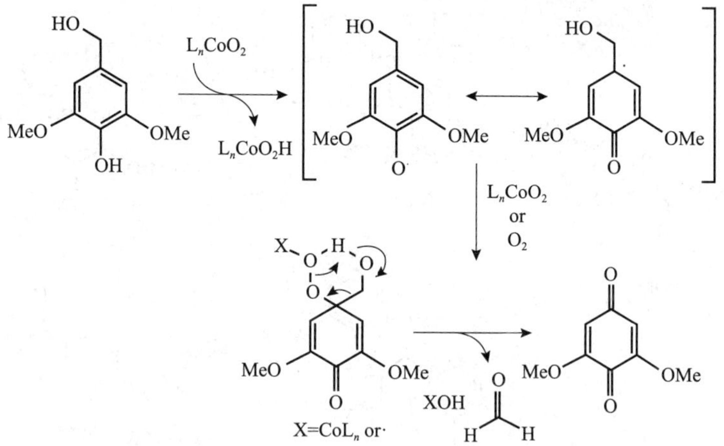

Figure 3 – 31 Proposed mechanism for the formation of quinone from the oxidation

ketone. Another interesting observation is that, when stoichiometric amount of oxidant CuOTf/TEMPO was used, the major products were quinone and aldehyde. The difference between catalytic and stoichiometric oxidation for product distribution is still unclear.

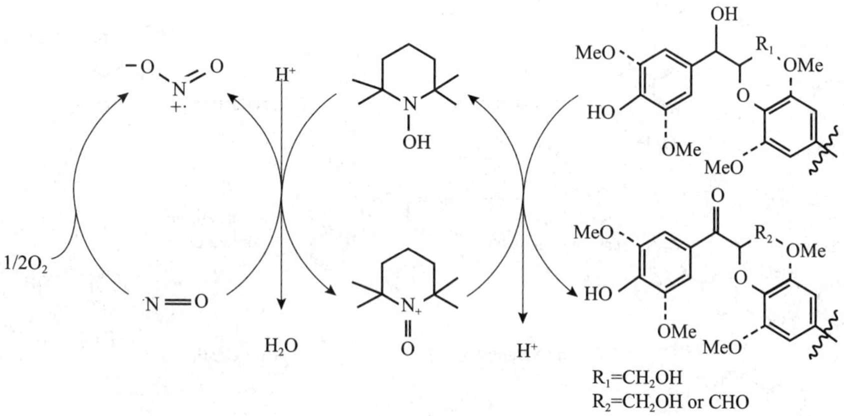

Figure 3 – 32 Mechanism of TEMPO oxidation C_4—C_β bond

Secondary depolymerization has been proposed for further degradation of post-benzylically oxidized lignin (Figure 3 – 33). Various oxidation techniques can be applied including Enol ether formation/hydrolysis, Zn reduction, Baeyer-Villiger oxidation/hydrolysis, Dakin oxidation, Beckmann rearrangement/hydrolysis, and Cu-ligand oxidation. Despite the good efficiency for those reaction candidates, phenols generated tend to polymerize under oxidative environment producing intractable mixtures of products. It is found that, if oxidative processing is conducted prior to lignin extraction, it is conceivable that conjugate esters already present in lignin materials could act as acid catalyst for hydrogenolysis of enol ether under thermochemical conditions (Figure 3 – 34). Overall secondary depolymerization is beneficial to obtain good monomer yield.

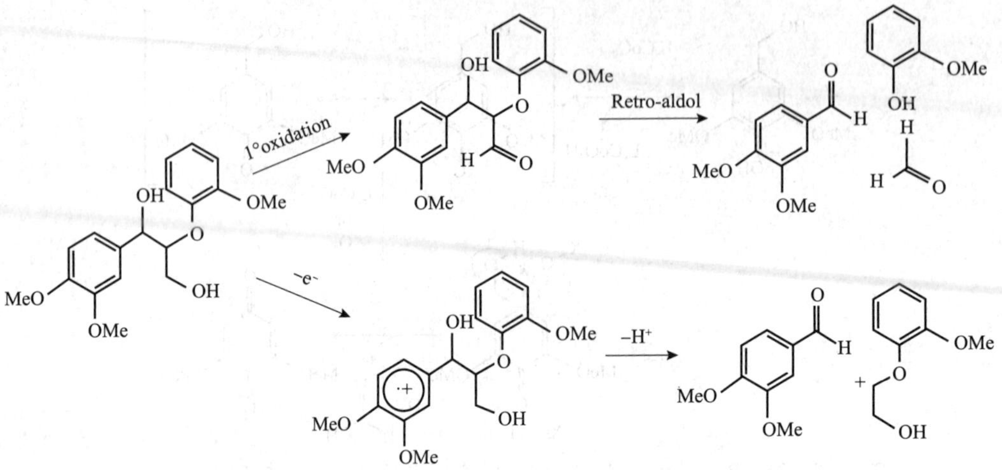

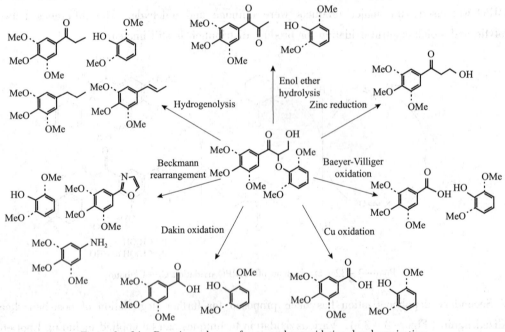

Figure 3-33 Possible reaction pathways for cleavage of C_α—C_β bond

Figure 3-34 Reaction schemes for secondary oxidative depolymerization

3.4.3 Lignin for fuels and chemicals

(1) Pyrolysis of lignin

As already mentioned, fast pyrolysis of lignin is a promising technology for producing bio-oil from lignocellulosic biomass. Pyrolysis of woody biomass usually consists two steps, one involving cracking of cellulose, hemicellulose and lignin *via* cleavage of covalent bonds, and then recombination of generated free radical compounds to form pyrolysis products. Although β-O-4 bond is the most dominant one in lignin structures, the exact mechanism still remains unclear at current stage. This is because structures of lignin are too complicated for researchers to isolate reaction intermediate or control kinetics of one particular reaction. Researchers have found model compounds to elucidate possible reaction mechanism. A β-O-4 type lignin model dimer has been investigated using GC—MS technique. Homolysis of C_β—O bond occurs at low temperature (300℃). Pyrolysis reaction is believed to initiate with C_β—O bond cleavage, owing to relatively low dissociation energy (221.4 kJ/mol, Figure 3 – 35), producing 4-methoxystyrene and guaiacol. Homolysis of C_β—O and C_α—O concerted decomposition mainly control the degradation process at a moderate pyrolysis temperature (500℃), producing 4-methoxystyrene, guaiacol, and carbonyl-containing substances.

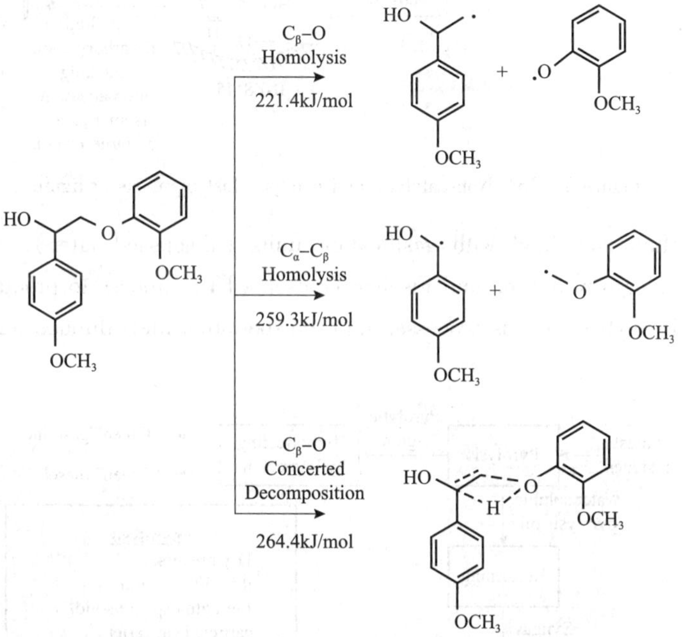

Figure 3 – 35 Mechanistic study on conversion of lignin model dimer

Various factors may affect the conversion of lignin during pyrolysis process. Reaction temperature plays a critical role in product distribution. Commonly, an optimal yield of bio-oil from wood can be obtained at temperatures of 450 ~ 500℃. Further increasing pyrolysis temperature up to 560℃ leads to decreasing char yield to 47%. Bio-oil yield is increased to a maximum value (57.1 % (wt)) at 500℃ and then decreased to 53.3% (wt) at 800℃. The maximum concentration of phenols (G-type and S-type phenols, p-methylguaiacol, vanillin) in bio-oil was as high as 79.3% obtained at 600℃.

From Figure 3 – 36, it is seen that, non-catalytic lignin pyrolysis (thermal conversion) eventually leads to the formation of coke. However, in the presence of porous materials, phenol compounds can be obtained after C_β—O bond cleavage. However, deoxygenation reactions occur in the presence of strong acidic catalysts such as ZSM-5. Acidic sites inside pore wall facilitate cleavage of C—O bond and generate aromatic hydrocarbons.

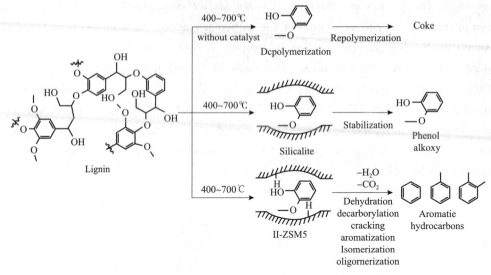

Figure 3-36 Non-catalytic and catalytic fast pyrolysis of lignin

Pyrolysis is often combined with gasification units (discussed later). The water soluble pyrolysis oil can undergo reforming and Fischer-Tropsch (FT) process to produce higher alcohols (Figure 3-37). Pyrolytic lignin is processed into transportation fuels through further hydrotreating steps.

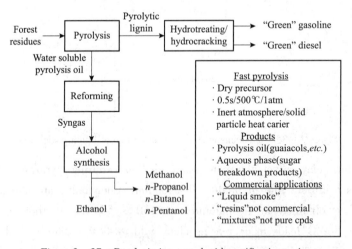

Figure 3-37 Pyrolysis integrated with gasification units

(2) Lignin for fuels and chemicals

As mentioned in pyrolysis part, aromatic compounds can be produced in the presence of acidic catalytic materials. Lignin has always been considered as a promising source for synthesis of bio-aromatics, mainly owing to continuous demand of high volume aromatics and derivatives. The basic units for lignin consist of phenol like functionalized parts (Figure 3-38), thus deoxygenation reactions would make lignin materials perfect candidates for aromatics.

National Renewable Energy Laboratory (NREL) has developed a multiple-step process to convert lignin into branched aromatic hydrocarbons that can be used as a blending component for

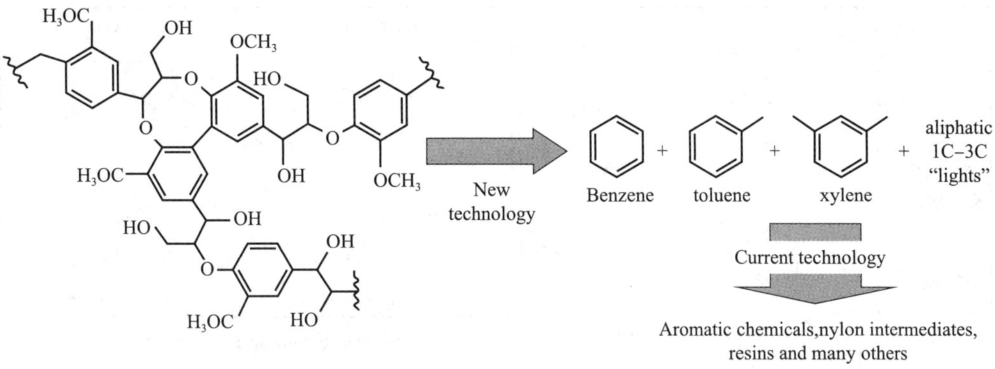

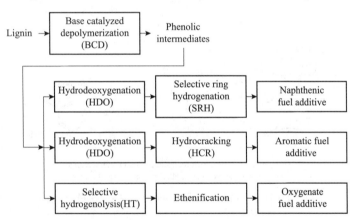

Figure 3-38 Lignin for aromatic chemicals

reformulated gasoline (Figure 3-39). The first step involves base-catalyzed depolymerization, breaking the C—O—C bonds in linkage, into phenolic intermediates. They can be further hydrotreated through hydrodeoxygenation and hydrocracking to give final gasoline-blending components. The final products often consist of naphthenic and aromatic hydrocarbons. This process is now at demonstration scale. Further technical and economic analysis should be conducted and challenges should be addressed for eventual industrial implementation.

Figure 3-39 Hydroliquefication of lignin feedstocks into fuel compounds

Since lignin is already oxygen functionalized material, it can be easily transformed into various oxygenates. Typical examples of downstream products include BPA, cyclohexanol, cyclohexanone, and nitrophenols (Figure 3-40).

Conventional gasification technologies have also been applied in transforming lignin feedstocks to oxygenate products. Gasification of lignin produces syngas. Subsequent water gas shift reaction further allows tunable H_2/CO ratio. H_2 can be used in fuel cells, hydrogenation process and power generation. Syngas is able to further generate methanol as a key intermediate in energy industry. Methanol can be further converted into dimethyl ether and olefins. Therefore, it is seen that lignin-derived syngas production units are readily implemented in industry. Besides Fischer-Tropsch (FT) technology to produce green gasoline and diesel, mixed alcohols and waxes can also be synthesized using lignin-derived syngas.

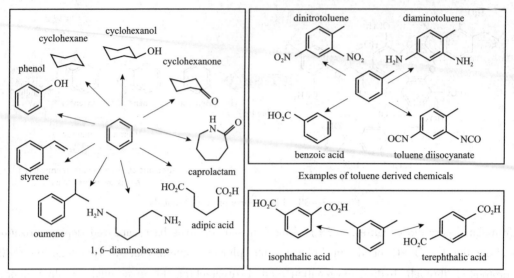

Figure 3-40 Conversion of BTX into downstream fuels and chemicals

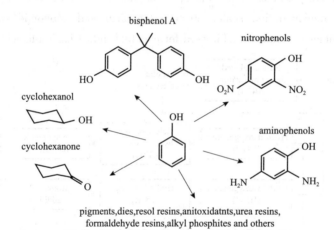

Figure 3-41 Conversion of phenols to value-added chemicals

Gasification technologies can treat lignin or mixed cellulose, hemicellulose and lignin feedstocks. A stand-alone gasifier unit can be directly implemented for converting forest residues (Figures 3-42, 3-43). This type of unit can also be combined with bioethanol production plant to fully utilize carbon feedstocks in biomass.

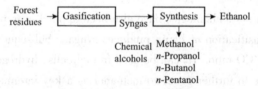

Figure 3-42 Conversion of forest residues to alcohols

Lignin is still considered as a virgin land for biomass conversion, as most current upgrading technologies still involves pretreating and separation processes (Table 3-4). It is still

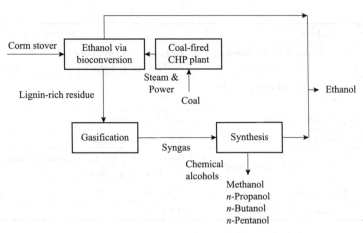

Figure 3-43 Conversion of corn stover to alcohols

economically challenging to obtain one or two particular product with large scale applications. Despite extensive efforts have been paid to convert lignin into structural materials, lignin materials are also regarded as promising feedstocks for high quality fuels and chemicals.

Table 3-4 Challenges for lignin conversion for power and fuels

Process	Products	Challenges
Combustion	Heat	Engineering and system management
Gasification	MeOH/DME, olefins	Gas purification Process scale up
Gasification	F-T fuels	Gas purification Catalyst durability Process scale up
Gasification	Mixed alcohols	Gas purification Catalyst durability Selectivity for C_{2+} products
Pyrolysis	Bio-oil	Stability of oil Transportation to petro-refinery
Fuels	Gasoline compounds	Process demonstration Catalyst life Process scale up

A number of fundamental research publications have been reported for conversion of lignin into chemicals, as chemical products are considered to contribute more significantly in economics of lignin utilization. Table 3-5 summarizes key downstream chemicals and relevant technologies involved in transformation technologies.

Table 3-5 Products from lignin and relevant reactions

Products	Technologies
BTX	Dehydroxylation, demethoxylation, dealkylation
Phenols	Dehydroxylation, demethoxylation dealkylation, hydrogenolysis

Continued

Products	Technologies
Lignin monomers: prpylphenol, eugenol, syringols, aryl ethers, alkylated methyl aryl ethers	Hydrogenolysis, hydrotreating dealkylation, hydrolysis
Oxidized lignin monomers: Syringaldehyde, vanillin, vanillic acid	Oxidation
Diacids and aromatic diacids	Oxidation, carbonylation
Adipic acid, aliphatic acids, polyols	Biocatalysis, oxidation
Aromatic polyols: cresols, catechols, resorsinols	Hydroxylation, dehydration, dealkoxylations reductions, ring saturation
Cyclohexanes	Reduction
Quinones	Oxidation

(3) Lignin for materials

We have witnessed increasing attention in academia, that depolymerization of lignin for fuels and chemicals are receiving wider interests, where novel solvents, catalytic materials and processes have been proposed and validated at laboratory levels. Actually, except for combustion and production of synthetic vanillin, the majority of downstream applications for lignin is targeted at dispersants, emulsifiers, binders, which account for nearly 75% of lignin applications. Lignin in those applications actually undergoes little or slight structural or surface modification, mainly representing low values and limited volume growth annually.

Main uses for lignin materials include carbon fibers, polymer modifiers, adhesives and resins. The major issues with lignin applications is that, raw lignin materials and isolated intermediates differ significantly in various properties such as reactivity, molecular weight distributions, melting points and polyelectrolyte features. Therefore, R&D should clarify the key factors that could alleviate the unfavorable complications in chemical and physical properties due to different biomass sources and processing conditions.

Lignin materials are found to be cost-effective precursors for replacing synthetic polymers such as polyacrylonitrile in the production of carbon fiber (Figure 3-44). Transforming 10% lignin could produce sufficient carbon fiber products to replace almost 50% of the steel use in passenger vehicles. Processing lignin materials to displace conventional carbon fiber products could reduce the sensitivity towards oil price, also favorable lowering environmental impact. Reduction of total car weight means better fuel economy.

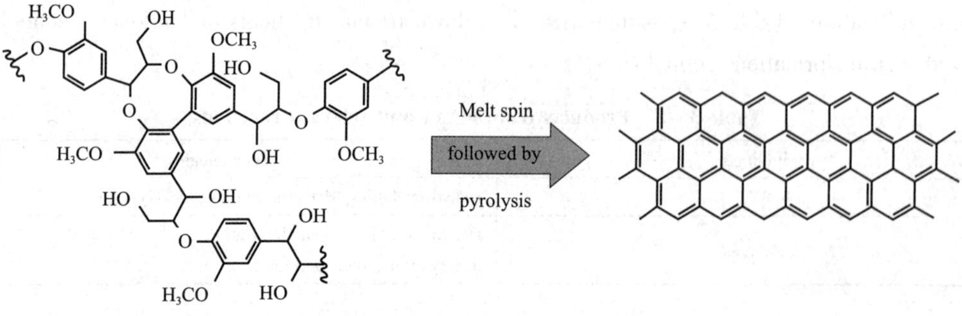

Figure 3-44 Lignin to carbon fiber

Practically, carbon fibers are manufactured *via* electrospinning of carbon precursors. Carbon fibers can be applied in lightweight composites, electromagnetic shielding, biomedical materials, batteries and supercapacitors. Production processes general include three main procedures. Fibers are spinning and under stabilization at 200 ~ 300℃, followed by carbonization under inert atmosphere at 1000 ~ 2000℃ or even harsher temperature. Lignin can be used as precursors, particularly for stabilization step, a thermoplastic spun lignin can be transformed into polymer matrix *via* crosslinking, oxidation and cyclization reactions to prevent fusion during subsequent thermal treatment. Stabilization step can be conducted under O_2, NO, SO_2, SO_3 and acidic conditions. Three commonly used carbon precursors have been used for commercial production, polyacrylonitrile, oil-derived pitch and regenerated cellulose. The main reaction pathways have been proposed in Figure 3 – 45.

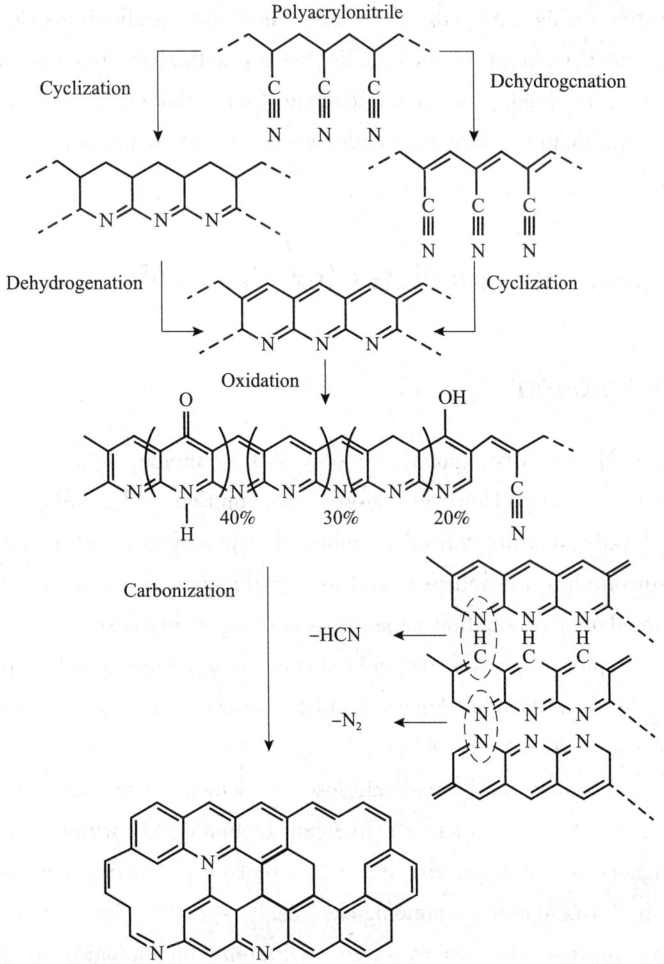

Figure 3 – 45 Conversion pathways for lignin to carbon fibers

A technical bottleneck for processing lignin-derived carbon is efficient and low-cost purification of lignin to remove sugars, salts and particulate impurities. Another technical challenge is to control lignin weight polydispersity. Thermal, melt flow and melting point should also be well dealt with for product quality.

Xu and colleagues have recently summarized in literature that, the production of carbon fibers from lignin can be divided into three categories: (a) carbon fibers from raw lignin without any further modification, (b) carbon fibers from physical lignin/ polymer blends, and (c) carbon fiber from modified lignin.

Mechanism of fiber formation from raw lignin was proposed as follow: degradation of linkages of lignin or aliphatic side-chain groups on the structure, demethoxylation reaction to release oxygen content, and formation of carboxyl and carbonyl groups induced by crosslinking of the oxidized lignin macromolecules. Carbonization step could further reduce the content of carboxyl and carbonyl groups and increase the percentage of aryl and condensed aryl carbons. Alternatively, carbon fibers can be produced by blending lignins with polymers, such as polylactic acid, polyethylene oxide.

Formaldehyde-free resins and adhesives are important applications for lignin materials. Technical challenges need to be resolved for effective, practical approaches for molecular weight and viscosity control, functional group enhancement (carbonylation, carboxylation, amination, epoxidation and de-etherification, that is, methoxy conversion to phenolic) to improve oxidative and thermal durability.

3.5 Platform intermediates for chemicals

3.5.1 Background

Since the start of 21st century, global economy and technology industries have witnessed the rapid growth in biofuel volume. However, global consumption of biofuels has almost reached a plateau in the past decade, mainly caused by minimal deployment of advanced biofuel routes and policy uncertainty surrounding the indirect land use by biomass. In sharp contrast, investment in biochemicals and biopolymer production capacity is starting to increase significantly, despite from a much lower base. Rapid progresses on technological development and facility implementation have been gained in North and South America and EU areas, where several processes use readily available sugars and starch feedstocks.

Pretreatment of cellulose and hemicellulose feedstocks has been developed to near-commercialization stage. This makes sugar conversion technologies, which can made from various woody, waste and lignocellulose materials. It is a fact that most existing sugar conversion pathways are realized through fermentation technologies, each specific downstream product can be synthesized following particularly design route. Different technologies at different stages of commercialization are required. Therefore, product conversion mapping is necessary for integrated biorefinery concept, which could an important role in sugar platform.

In both literature and industry, scattered examples are available. Comprehensive studies and assessment have conducted to screen the possibility of technological development for potential valuable products in downstream side.

Figure 3-46 presents the basic feedstocks for sugar routes, including glucose, fructose, galactose, xylose, arabinose, ribose, lactose, sucrose and maltose. Reforming reactions can transform those sugars, or even sugar-derived polyols into H_2, alcohols and aromatics. Dehydration and oxidation are another two important reactions, which can upgrade sugars into various acids, polyols and ketones, examples of which include levulinic acid, hydroxylmethylfurfural, sorbitol, glycerol, *etc*. Based on the streams, the dominant value-added products are still derived from biological processes.

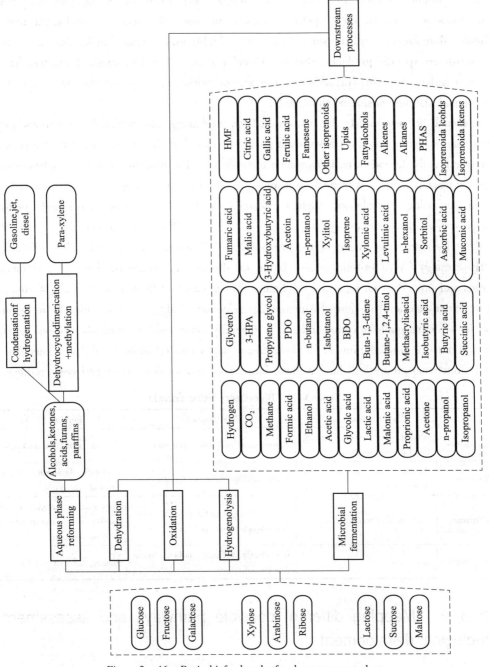

Figure 3-46 Basic biofeedstocks for downstream products

United States and EU both identified key intermediate compounds, often referred as platform chemicals/intermediates for biochemicals. EU reported ethanol, lactic acid, propanol, glycerol, succinic acid, isobutanol, xylose/arabinose, levulinic acid, glucose/fructose as basic feedstocks for different downstream chemicals. U. S. Department of Energy published reports covering 12 building blocks including succinic acid, furan dicarboxylic acid, 3-hydroxyl propionic acid, aspartic acid, glucaric acid, glutamic acid, itaconic acid, levulinic acid, 3-hydroxybutyrolactone, glycerol, sorbitol, xylitol/arabinitol. Nevertheless, the examination of potential platform intermediates is based on their upstream sugars and both chemical and biological routes for synthesis. More importantly, transformation of those highly functionalized molecules into more than 200 downstream specific products, due to molecular similarity and easiness of reaction has been taken into serious account. In addition, the known market size, properties, performances and prices should also be considered thoroughly.

It should be noted that, EU's criterial on evaluating the potentials of various platform intermediates is purely based on downstream products. Therefore, it can be seen that, the conversion mapping is heavily dependent on fermentation technologies. The potential of actual implementation of production plant is defined as technology readiness level, where nine stages have been referred: basic research, technology formulation, applied research, small scale prototype, large scale prototype, prototype system, demonstration system, commercial system, full commercialization.

U. S. Department of Energy screened more than 300 potential compounds prior to finalizing to 12. The strategy for fit criteria is also based on downstream products. However, estimates of materials and properties for potential candidates are summarized from over 75 years of cumulative industry experience and for downstream product selection. Table 3 – 6 provides basic evaluation factors for selecting platform compounds. It is generally found that, compounds with one functional group will have a limited number of downstream applications, while candidate molecules with multiple functional groups will have a much larger potential for derivatives and new products.

Table 3 – 6 Biorefinery fitted criteria

	Product replacement	New products	Building blocks
Characteristics	Advantageous over existing fossil derived products	Improvement and new applications	Diverse portfolio of products
Examples	Acrylics from propylene or lactic acid	Lactic acid from glucose fermentation	Succinate, levulinate, glutamate, glycerol, etc.
Up Streams	Large market volume	Low cost, no competitive fossil routes, new properties and applications	New starting material to reduce market risk, incorporate with existing products
Down Streams	Lower cost	Not clearly defined market, future commercialization	Identify hot spot in R&D

3.5.2 Mapping different possible pathways and assessment of technological development

One should keep in mind that, the ultimate goal of replacing fossils with bio-derived products is to produce fuels and chemicals with similar structures as derived from petroleum and coals. Therefore, the basic products manufactured from fossil fuels need to be identified prior to detailed analysis of each important platform compound. Benzene, xylene, toluene and butanes are main platform intermediates from petroleum (Figure 3-47). Catalytic transformation of the four compounds to ethylbenzene, cyclohexane, cumene, butylene and butadiene provides megaton intermediates for end applications.

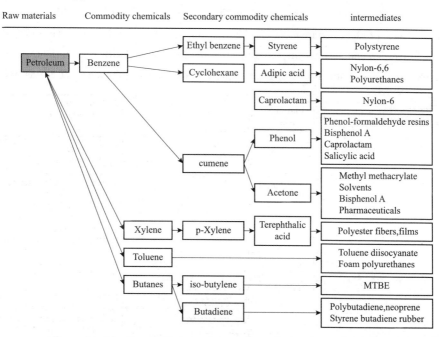

Figure 3-47 Conversion route for petroleum feedstocks to chemicals

In particular, benzene is transformed into ethyl benzene for further dehydrogenation applications for manufacturing styrene, which is an important monomer for polystyrene. Hydrogenation of benzene leads to the formation of cyclohexane. Homogeneous oxidation of cyclohexane gives cyclohexanone and adipic acid as main products, the mixture of intermediate often referred as KA oil. Adipic acid and derivatives are used as precursors for Nylon-6, 6 and polyurethanes. Another derivative from cyclohexane is caprolactam, a critical starting material for Nylon-6 product.

Cumene is converted to phenol and acetone as megaton products. A variety of everyday end products, such as phenol-formaldehyde resins, bisphenol A, salicylic acid, for solvent and pharmaceutical industries, can be obtained from cumene.

A well-known product from xylene is terephthalic acid. Terephthalic acid reacts with ethylene glycol to yield poly ethylene terephthalate (PET) as final product. PET is almost used everywhere in our life, particularly for plastic bottles and containers.

In this map, toluene can be converted to produce toluene diisocyanate. Isobutylene has been widely for MTBE production for performance enhancer of gasoline. Butadiene is transformed *via* polymerization to poly butadiene, neoprene and styrene-based rubber products.

Natural gas, mainly consisting of $C_1 \sim C_5$ alkanes, is another essential feedstock for various chemicals and materials (Figure 3-48). Specifically, methane is mainly burned for power generation and heating requirement. Oxidation of methane gives formaldehyde and methanol, CO and H_2 as main products, which are essential building blocks for olefins and oxochemicals. Ethane and propylene can undergo dehydrogenation in cracker reactors to generate ethylene and propylene products. Ethylene and propylene have been widely used for polymers. Epoxidation of those light olefins yield ethylene epoxide and propylene epoxide. Hydrolysis of epoxides leads to the generation of glycols, which as essential components for antifreeze, resins, fibers, *etc*. Chlorine is important to diverse natural gas products. Vinyl chloride is starting material for polyvinyl chloride (PVC). CO and H_2 mixtures, commonly known as syngas, can be catalytically transformed formaldehyde, MTBE, acetic acid and ammonia. Those are typical performance enhancers for resins and fuels.

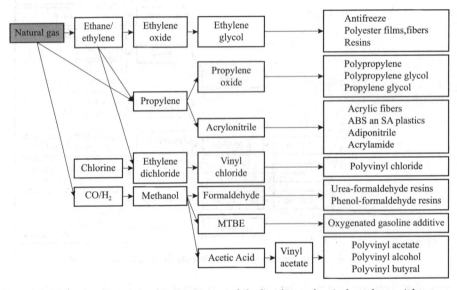

Figure 3-48 Conversion route for natural feedstocks to chemicals and materials

Petroleum-based conversion map has been established for over 80 years in chemical industry. The general methodology in transforming oil to chemicals, is to functionalizing molecules with oxygen or other heteroatoms, such as Cl, N, S, *etc*. Different from hydrocarbons, biomass, particularly for cellulose and hemicellulose, is highly functionalized or oxygenated hydrocarbons. Thus an alternative methodology is often employed, namely defunctionalization. The philosophy for selecting platform compounds have been discussed, while EU and U.S. actually proposed different intermediates for conversion map/route.

3.5.3 Platform compounds from land biomass

EU selected acrylic acid, adipic acid, acetic acid, 1,4-butanediol, ethanol, 3-hydroxyl propionic acid, succinic acid, lactic acid, butanol, iso-butene, isoprene, farnsene, propanediol, *p*-xylene, sorbitol, xylitol, furan dicarboxylic acid, 5-hydroxyl methylfurfural, levulinic acid, furfural, itaconic acid, ethylene, poly-hydroxyl alkanoates, ethylene glycol and algal lipids as platform compounds. They have classified the conversion technology for those compounds for various stages, research, pilot, demonstration and commercialization, the number of which represents the TRL stages (Figure 3 - 49). The TRL stages for 25 platform compounds are different owing to existing and emerging markets and technologies. For example, bioethanol is still the dominant sugar-based product. Markets for butanol, acetic acid, lactic acid are also large. Those chemicals can potentially replace petroleum-based counterpart. Xylitol, sorbitol and furfural also display significant market but without competitive petrochemical alternatives. The competitiveness edge for 3-hydroxyl propionic acid, acrylic acid, isoprene, adipic acid and 5-hydroxyl methylfurfural is still small compared with fossil-derived counterparts, thus new technologies are required for those bio-based chemicals to be more acceptable for downstream consumers.

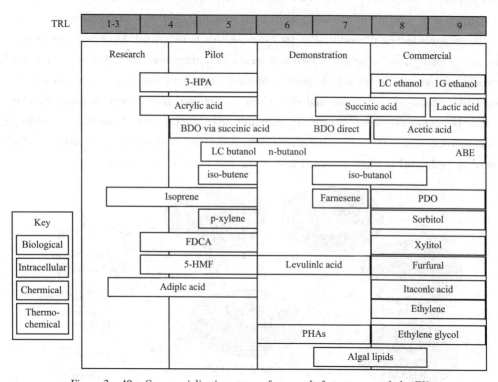

Figure 3 - 49 Commercialization status of sugar platform compounds by EU

U. S. Department of energy has classified platform compounds based on carbon number. The technological development status has also been discussed and compared with potential competitive counterparts. Actually more than 30 platform chemicals were initially selected, according to carbon number. C_3 compounds include glycerol, lactic acid, 3-hydroxyl propionic acid, propionic acid, malonic acid and serine. C_4 include succinic acid, fumaric acid, malic acid, aspartic acid, 3-hydroxy-butyrolactone, acetoin and threonine. C_5 molecules are itaconic acid, furfural, levulinic acid, glutamic acid, xylonic acid and xylitol/arabitol. C_6 intermediates are citric acid, 5-hydroxyl methylfurfural, lysine, gluconic acid, glucaric acid and sorbitol. Gallic acid and ferulic acid are aromatic acids which is derived from lignin materials.

The 30 key intermediates were eventually narrowed down to 12, which have already been mentioned in previous sections. In the following section, selected platform compounds from EU and U. S. will be discussed in terms of synthetic route, main products and derivatives.

(1) Succinic/Fumaric and Malic acids

At present, no chemical pathways have been developed for synthesis of C_4 dicarboxylic acids. The main technical barriers for this pathway include reduction of acetic acid by-product and increasing overall productivity. Most typical downstream products are tetrahydrofuran, 1,4-butanediol, γ-butyrolactone for hydrogenation process. Pyrrolidinone family products can be obtained by reductive amination chemistry.

Low cost fermentation is still a grand challenge for production, while productivity can be further improved to 3% (*wt*) to be economically competitive. Another issue is with pH values, since formation of succinic acid would significantly enhance acidity of reaction medium.

The only real technical consideration for hydrogenation route, is the development of catalysts that would not be affected by impurities formed and added in fermentation. This is a significant challenge but would not necessarily be a high priority research item until the costs of the fermentation are substantially reduced. Pyrrolidinone and derivatives are obtained from γ-butyrolactone. Succinic acid can also be converted more directly to pyrrolidinones through the fermentative production of diammonium succinate. The main advantage is that, conversion of succinate to pyrrolidones eliminates the need to modify pH in direction production of succinic acid (Figure 3-50).

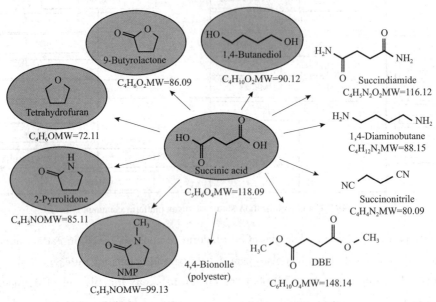

Figure 3-50 Succinic acid and derivatives

(2) Furan dicarboxylic acid (FDCA)

FDCA is a member of furan family, which is synthesized through dehydration and oxidation of fructose. Oxidation reaction can occur using molecular O_2 or electrochemical oxidation. This process is conducted using 5-hydroxyl methylfurfural as the starting material, which is the dehydration product from C_6 sugars. There is no report published on biological conversion of 5-

hydroxyl methylfurfural to FDCA, as catalysts for chemical oxidation have already shown remarkable activity in either alkaline or base-free medium, despite some improvement on selectivity can still be required. Noble metal catalysts are predominantly used for this reaction. Earlier days this reaction was carried out in alkaline medium, where generation of waste salts was a main issue. Nowadays, oxidation of 5-hydroxyl methylfurfural to FDCA can be easily realized using hybridized noble and non-noble metal catalysts. Therefore, cost of 5-hydroxyl methylfurfural is the main limiting factor for further implementation of this technology.

FDCA is regarded as an emerging platform intermediate because it is considered as the most promising alternative to replace terephthalic acid and its polymer, PET materials. FDCA market is projected to grow at a significant rate, with a market volume up to 560,000 t/a by 2025. Replacement of PET is expected to be approximately 322,000 tons. The cost for producing 5-hydroxyl methylfurfural is still unfavorably high, thus the price of FDCA is economically less competitive compared with conventional petro-based products. PET has a market size approaching 4 billion lb/a. The price for PET is in the range of $1.00 ~ 3.00/lb for uses as films and thermoplastic engineering polymers.

(3) Glucaric acid

Glucaric acid is an important member for oxidized sugars (Figure 3-51). Those chemicals represent a huge future market in fine chemical industry. Oxidation of glucose to glucaric acid is known to reach almost 80% yield in the presence of Pt-based catalysts in aqueous medium. But existing technologies for glucaric acid synthesis are still highly dependent on nitric acid chemistry. In particular, concentrated nitric acid functionalizes with sulfuric acid to oxidize glucose in aqueous medium, which occur within a few minutes, causing temperature rise from room temperature to almost 200℃. Different co-products with shorter carbon numbers have been generated with almost 45% yield. Development of efficient processes for glucaric acid synthesis is also important for converting xylose and arabinose to relevant dicarboxylic acids.

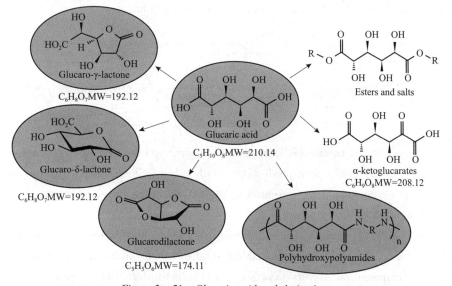

Figure 3-51 Glucaric acid and derivatives

Selection of glucaric acid is established as it is the starting materials for various highly functionalized products. One of the most well-known products is adipic acid, which is used for manufacture of nylon-6,6. Hydrodeoxygenation of glucaric acid can give adipic acid as the main product. Other downstream derivatives include lactone, polyamides and esters.

The main technical barrier for glucaric acid synthesis is the selectivity, eliminating the need for nitric acid, using molecular O_2 as the green oxidant. Further technical barriers include development of selective methods for sugar dehydration to transform glucaric into sugar lactones, particularly glucaric dilactone, an analog of isosorbide.

(4) Adipic acid

It is a C_6 linear chain dicarboxylic acid often used as monomers for nylon and polyurethane. Approximately 90% of adipic acid worldwide has been applied for production of nylon-6,6, a specialty product with high performance in resins, which can be further transformed into fibers for carpeting, automobile tire cord and clothing. Adipic acid is also used to manufacture plasticizers and lubricant components. Food grade adipic acid is used as a gelling aid, an acidulant, and as a leavening and buffering agent.

At present, adipic acid is produced through petrochemical route, namely as oxidation of cyclohexane, benzene or phenol in a two-step process. About 93% adipic acid is synthesized from oxidation of cyclohexane. This reaction produces KA oil as main product (cyclohexanone and cyclohexanol), followed by stoichiometric oxidation by nitric acid to eventually form adipic acid. Environmental issues related with fossil-derived adipic acid are very serious, blocking further implementation of conventional technologies. Notably, N_2O (greenhouse gas effect 300 times worse than CO_2), is a by-product in the petrochemical processes, leading to greenhouse gas emission approximately 60 kg N_2O/ton adipic acid.

As mentioned in glucaric acid section, hydrodeoxygenation reactions eliminating internal-OH groups can transform glucaric acid into adipic acid (Figure 3-52). Biological routes also include conversion of muconic acid and glucaric acid in aqueous medium.

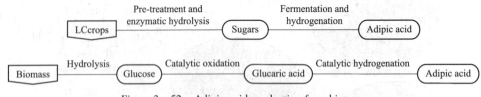

Figure 3-52 Adipic acid production from biomass

Bio-adipic acid is not commercially available at current status. Only a pilot scale of 1 t/a is established in U.S. Compared with fossil-adipic acid with a 2.7 million metric tons in volume, bio-adipic acid is expected to expand to around 10,000 t/a in the next few years.

(5) Levulinic acid

The critical role of levulinic acid in the conversion routes for biomass has already been discussed in details (Figure 3-53). Chemical synthesis of levulinic acid is dependent on the technological development on conversion of 5-hydroxyl methylfurfural. Decomposition of fructose or

5-hydroxyl methylfurfural directly give levulinic acid and formic acid as major products. Technical barriers in this area is development of efficient heterogeneous catalysts for degradation reactions to replace homogeneous sulfuric acid.

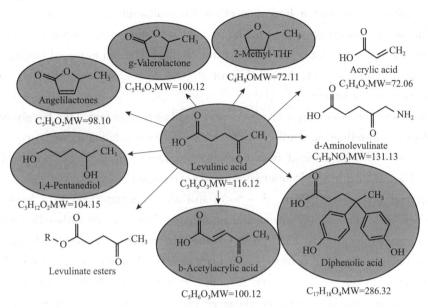

Figure 3-53 Levulinic acid and derivatives

The main downstream derivatives for levulinic acid include methyl tetrahydrofuran, γ-valerolactone, acetyl acrylates, and diphenolic acid. It is necessary to mention that, levulinate esters are promising additives for fuel performance enhancers. But current research attention is still focused on activation of C=O bond. In this area, selective reduction of carboxylic and ketonic groups into alcoholic ones is the main challenge. For example, hydrogenation of levulinic acid easily form γ-valerolactone with good selectivity (>99%). However, further hydrogenation for ring-opening reaction is inhibited by poor catalytic activity of solid catalysts.

For specific products, methyl tetrahydrofuran is also a valuable fuel additive. Aminolevulinic acid is a herbicide with target market volume being 200~300 million lb/a. Production of δ-aminolevulinic acid and β-acetylacrylic acid is important for manufacturing new acrylate polymers. Complete reduction of levulinic acid to 1,4-pentanediol is also an important application for production of new polyesters.

In this conversion route, cost-effective production of levulinic acid is still the bottlenecking factor for further development of derivatives. Overall yield for levulinic acid is approximately 70%. Besides reduction, oxidation of levulinic acid can also produce succinic acid and acrylic acid, using cheaper oxidant could be another topic of great interests to industry.

(6) Glycerol, sorbitol and xylitol

Sugar derived polyols are among mostly investigated platform compounds in the area of biomass conversion. Glycerol has already been well-established in chemical industry with annual global production over 1 million tons. Glycerol and esters such as glycerol stearate, oleate are currently produced in large scale. Glycerol is produced naturally from vegetable oil and animal fat. The most well-known transesterification of vegetable oil produce fatty acid methyl/ethyl esters, and glycerol as by-production. The prices for glycerol actually fluctuate with crude oil. Development of a biodiesel market could have a huge impact on the availability and use of glycerol. This is because that, low petroleum price prevents larger implementation of bio-diesel production plants, which affects the sufficient supply of crude glycerol. Refined glycerol is priced ranging from $0.6 ~ $1.0/lb in current market, however, most glycerol derivatives have relatively smaller and more fragmented market volume.

Market growth for bio-diesel and government regulations are important for glycerol as key building block of bio-refineries. Particularly for EU zone, transportation diesel is required to have large fraction of feedstocks derived from renewable resources. In U.S. however, only a small fraction (~4%) of total on-road diesel is replaced by bio-diesel.

For glycerol derivatives (Figure 3-54), oxidation reactions lead to a broad family of products which could serve as new chemical intermediates. The hydroxyl carboxylic derivatives are regarded as simple compounds with great promises in replacing conventional monomers for polyesters and nylons. Targeted polyesters have markets of 2~3 billion lb/a, at values between $~3.50/lb, while nylons are a 9 billion lb/a market with values higher than $2.3/lb. The technical barriers for manufacturing those chemicals include selective chemical oxidation catalysis. Simple and green oxidants such as molecular O_2 are more favorable.

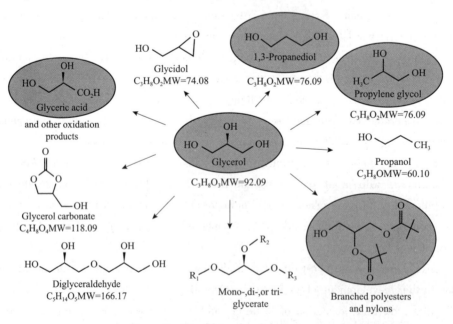

Figure 3-54 Glycerol and derivatives

Selective deoxygenation, namely hydrogenolysis technology leads to the formation of a number of important intermediates. 1,2-and 1,3-propanediols are critical derivatives for future polyester industry. And there is strong technological incentive to drive relevant reactions from laboratories to industry. Petrochemical routes for 1,2-propanediol have been established through propylene epoxidation-hydration and co-production during DMC synthesis. 1, 3-propanediol is currently produced through fermentation technologies using glucose or glycerol as starting materials. Development of chemical catalysis should be more cost-competitive with existing routes.

Sorbitol is the product from hydrogenation of glucose. Sorbitol production has been long commercially operated through hydrogenation of glucose with a global capacity of over 200 million lb/a. Ni based catalysts are widely used for hydrogenation processes under 100 ~ 120℃ in batch mode, to ensure complete conversion of glucose. This is very important, as the downstream applications for sorbitol in food industry are strict on the contention of reducing sugars. Main technical barriers for are focused on how to switching operation mode from batch to continuous one.

Isosorbide is among the most important derivatives from sorbitol (Figure 3 – 55), as inexpensive sorbitol is critically important for cheaper isosorbide. Isosorbide is a very effective monomer to raise the glass transition temperature of polymers. It is also regarded as an important alternative to replace BPA for plastic bottles, which show detrimental effect for brain development particular for infants. Technical challenges for producing isosorbide from sorbitol is product yield and durable dehydration catalytic systems. A yield of 90 + % would be more economically favorable for technological development. In the meantime, conversion of sorbitol into glycols such as ethylene glycol and propylene glycol (1, 2-propanediol) provides another way of producing those megaton intermediates. However, yields of glycols are not economically attractive compared with conventional routes. New catalysts with selectivity higher than 70% are preferable for commercialization of this reaction. Copolymerization with other glycols in the unsaturated polyester resin market would be a major opportunity for other applications for sorbitol.

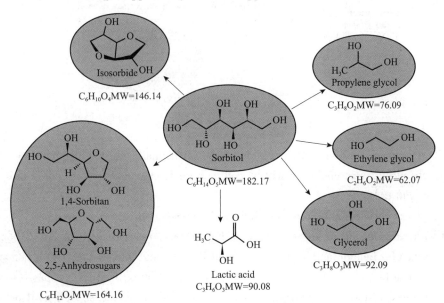

Figure 3 – 55 Sorbitol and derivatives

Similar to glycerol and sorbitol, C_5 polyols such as xylitol and arabinitol have great potentials for producing specialty chemicals for advanced materials. Production of xylitol and arabinitol is still heavily dependent on fermentation technologies. Biological catalysis offers both hydrolysis of hemicellulose and efficient reaction rate for hydrogenation of xylose and arabinose. There is limited commercial production of xylitol and no commercial production of arabinitol. Xylitol is widely used as non-nutritive sweetener. This process can be incorporated with glucose hydrogenation. Many transition metals including Ni, Ru and Rh can easily make this reaction achievable. There is no major technical barrier associated with the production of the five-carbon sugar alcohols xylitol and arabinitol.

For derivatives, several highly valuable products such as xylaric acid, and functionalized polymers (Figure 3 - 56). Xylaric acid requires active and selective oxidation chemistry for xylitol. This reaction has actually been demonstrated to be feasible for sorbitol oxidation to glucaric acid. In addition, more efforts need to be developed using molecular O_2 rather than nitric acid. Direct polymerization with other glycols for unsaturated polyester resin market would be one of the major opportunities.

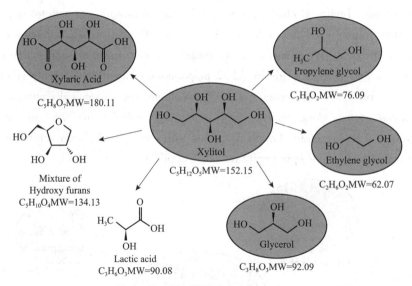

Figure 3 - 56 Xylitol and derivatives

In all, C_5 sugars such as xylose and arabinose are actually outstanding building blocks for commodity chemicals. One challenge will be getting a relatively clean feed stream of these sugars.

(7) Acrylic acid

It is a clear, colorless and corrosive liquid with characteristic acrid or tart smell. Acrylic acid can react with other monomers or even itself to form esters, *via* activation of C=C bond in the molecule. Examples include acrylamides, acrylonitrile, styrene, butadiene. Those homopolymers or copolymers can be used in a variety of products such as plastics, coatings, adhesives, diapers, fibers and textiles, resins, detergents and cleaners, synthetic rubbers, as well as floor polishes, and paints.

Conventional approaches producing acrylic acid is conducted from oxidation of propylene, which is derived from naphtha cracking (Figure 3 - 57). Major producers are BASF, Dow Chemical, Arkema, Nippon Shokubai, Jiansu Jurong Chemical, LG Chemical, Mitsubishi Chem, and Shanghai Huayi. Bio-acrylic acid can be produced from dehydration of 3-hydroxypropionic acid, which is obtained from sugar fermentation. Glycerol has also been demonstrated to give 3-hydroxypropionic acid through dehydration to acrolein and followed by oxidation reaction. Alternatively, lactic acid can also dehydrate to yield acrylic acid. Those bio-route is not commercially available yet.

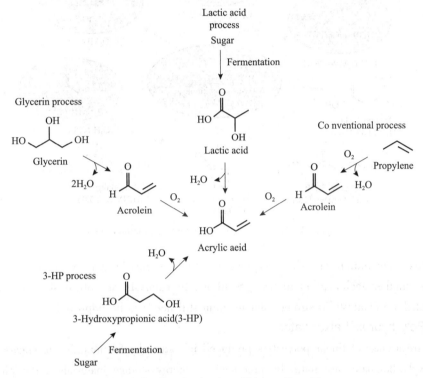

Figure 3-57 Production of bio-acrylic acid

The main downstream applications for acrylic acid is polyacrylic acid for superabsorbent polymers. Worldwide production of acrylic acid is approximately 5 million t/a with an estimated market value of $11 billion. The annual production volume for bio-based acrylic acid, still in pilot phase, is only around 300 tons.

(8) 3-Hydroxyl propionic acid

This is a C_3 organic acid with hydroxyl and carboxylic groups. Different from lactic acid, which has a-OH group at middle carbon, acrylic acid has a terminal-OH group at end carbon. At present, no relevant petrochemical technologies are available for production of 3-hydroxypropionic acid. The main technical challenge for producing 3-hydroxypropionic acid is the development of an organism with the appropriate pathways. A minimum productivity of 2.5 g/(L · h). needs to be achieved in order to economically competitive.

Downstream derivatives of 3-hydroxypropionic acid include 1,3-propanediol and acrylates (Figure 3-58). Technical barriers for conversion of 3-hydroxypropionic acid to 1,3-propanediol is the selectivity towards reduction of carboxylic acid under mild pressure and temperature. Robust solid catalysts sustaining inhibitory elements or components of biomass based feedstocks are required.

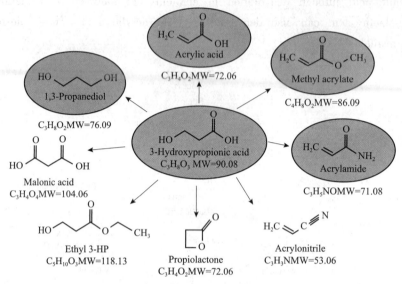

Figure 3-58 Derivatives of 3-hydroxypropionic acid

Selective dehydration of 3-hydroxypropionic acid to acrylic acid without significant side reactions is another technical challenge. Solid acidic catalysts are obviously preferred, while durability of those catalysts in strong polar medium still needs to be improved.

(9) Poly-hydroxyl alkanoates

They are a class of linear polyesters produced in nature by fermentation of sugars or lipids. Poly-hydroxyl alkanoates are naturally produced as energy storage materials under physiological stress condition. These plastics are biodegradable and can either be thermoplastic or elastomeric materials.

Polyhydroxybutyrate and polyhydroxyvalerate are common poly-hydroxyl alkanoates in nature. Several features for polyhydroxybutyrate, such as mechanical, physical and thermal properties are similar to many types of plastics polypropylene (PP), polyethylene (PE), low density polyethylene (LDPE), high density polyethylene (HDPE), polyvinylchloride (PVC), polystyrene (PS), and polyethylene terephthalate (PET). The global production of poly-hydroxyl alkanoates is estimated to be 54,000 tons at present. However, the process cost till need to be further improved, as the current market price for poly-hydroxyl alkanoates is around $4,400/t.

The primary value proposition of poly-hydroxyl alkanoates are their remarkable biodegradability under various conditions. They have good resistance to moisture, aroma barrier properties and can form clear films. A key advantage of such polymers is that, they can be processed in existing petrochemical plastic processing plants.

3.6 Fuels and chemicals from ocean biomass

While major research efforts have been put on land-based biomass, conversion of ocean biomass has also attracted wide interests. Around 6 ~ 8 million tons of crustacean shells are generated annually worldwide, the majority of which have been landfilled or dumped as fishery waste without full utilization. Direct combustion of those waste shells leads to release of significant amounts for CO_2 and NO_x compounds. Chitin is a type of biopolymer which account for 15% ~ 40% (wt) of crustacean shells (Figure 3 – 59). Chitin can be industrially obtained from crab and shrimp shells in white powder form, in multiple procedures including after demineralization, deproteination and decolorization. Chitin is considered as world's second most abundant bio-based polymer compared with cellulose with an estimated global production of 100 billion t/a. Chitin is structurally full of-OH and-NH_2 groups, making it perfect candidate for N containing materials for other applications.

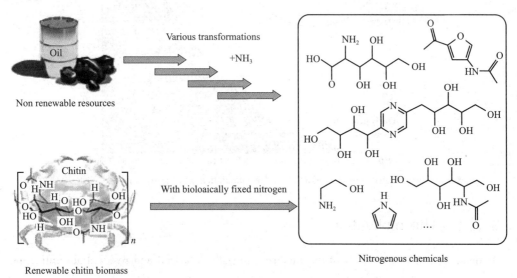

Figure 3 – 59 Petrochemical and chitin biomass for N-containing products

Various chitin related products such as 2-(acetylamino)-2-deoxy-d-glucose, glucosamine, chitosan and chitin polymer have been studied widely in the past decade. Those N containing chemicals and materials have been successfully obtained at laboratory level for amines, amides and heterocyclic compounds. The have been processed along with carbohydrate products *via* various reactions such as hydrogenation/hydrogenolysis, oxidation.

3.7 Global movement for biodegradable plastics

Global environmental crisis has motivated researchers worldwide to looking for alternative and sustainable plastics to substitute existing ones, which are mainly made from petroleum and coal

(Figure 3-60). To develop sustainable plastics, researchers are creating polymers that not only have excellent functional properties, but also offer reliable options for end-of-life management. In this context, life cycle assessment provides information to underpin these development, improving the efficiency of polymer production and reducing energy input. Considering the sources of carbon on earth, it may be beneficial to use biomass and CO_2 as raw materials to make new plastics.

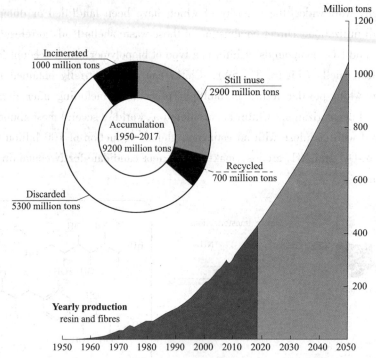

Figure 3-60 Global plastic production and accumulation and future trends (UNEP 2021)

3.7.1 The motivation

At present, the vast majority of plastics used in both industrial and residential applications is derived from petrochemicals. Continuing use of those raw materials will ultimately requires significant reduction in energy input and carbon footprint. Reduction of carbon emission demand considerably amounts of efforts in developing new technologies. In the long term, it seems that the world would undergo a gradual shift from a manufacture system predominately based on fossil fuels, to making new products from other resources including recycled chemicals, and renewable feedstocks.

As known to general public, most common petro-derived plastics are polymers such as polyethylene and polypropylene, which will take many years to finally degrade in natural environment. This is because those polymers do not have such unique chemical functional groups that could act as breaking points for chemical or biological degradation. These materials are often assembled to form highly crystallized morphology which confer remarkable thermal and mechanical properties, making them water resistant and extremely durable in and after uses.

Aquatic organisms are continuously exposed to litter and plastic pollution (Figure 3 – 61). Ongoing monitoring results have confirmed that physical collisions with macro-sized plastics by marine mammals, fish, birds, reptiles and plants are a direct source of fatalities. Microplastics are posing strong impact on marine life. Microplastics are continuously releasing a wide range of toxic chemicals, metals and micropollutants into open surface waters. Despite of unevenly studied mechanisms involved in microplastics, three mechanisms could be considered to have been "demonstrated": (a) inhibition of food assimilation and/or decreased nutritional value of food, (b) internal physical damage, and (c) external physical damage.

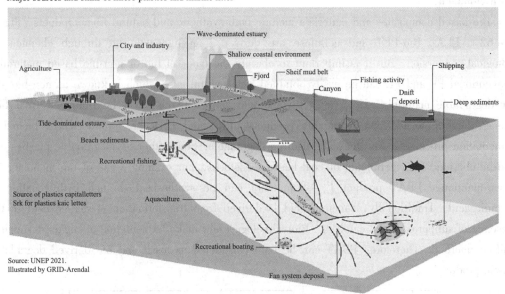

Figure 3 – 61 Major sources and sinks of microplastics and marine litter

More recently, the potential risk of **nanoplastics** in seafood has been raised by experimental studies. Compared to **microplastics**, nanoplastics show enhanced mobility in the tissues of living organisms. More critically, their larger surface to volume ratio increases the potential concentration of harmful chemicals they can adsorb. But the detailed mechanism is still unknown due to limited demonstration.

The chemicals found in microplastics include additives such as flame retardants, plasticizers, antioxidants, UV stabilizers, and pigments. Many additives are added intentionally to enhance plastic properties. Although those microplastics and nanoplastics are often chemically inert, the chemical additives may be harmful to human health. Some of the chemicals associated with plastics are recognized as mutagens and carcinogens.

Almost no commercial polymers are explicitly design for easy chemical or biological recycling. Only a handful examples are available for easy degradation in environment, including polyester polybutylene adipate terephthalate (PBAT), which is often used as packaging materials as they display similar properties with polypropylene. One important option to design degradable materials is to introducing oxygen-containing groups into hydrocarbon chains. Another alternative is to seek non-permanent interaction among chains to produce polymers which can dissemble easily.

3.7.2 Polymers from renewable resources

Bio-based plastics only account for a tiny portion of industrial polymers. Polylactic acid (PLA) is one of the most commonly used sustainable plastics that can be decomposed through biological degradation. Thanks to the unique alphatic ester groups in molecule, PLA can be easily broken down into small molecules such as CO_2, or even simple industrial composting processes. Life cycle assessment suggests that, PLA could yield very low and even negative carbon footprint in environment.

Bio-based composites and materials include both synthetic and natural fiber products (Figure 3-62). PLA, Bio-PET and cellulose are derived from lignocellulose through chemical or biological conversion. But it is important to classify the types of bio or non-bio based materials. Conventional PP and PE can be easily obtained from natural gas, which is known as non-bio and non-degradable materials. However, ethylene and propylene can also be produced from bio-syngas, leading to bio non-degradable products. PLA and PHA are produced *via* biological fermentation processes. And they are also bio-degradable. Some other bio-degradable plastics can also be obtained from petroleum or coal-based feedstocks. For example, one of the monomer for above-mentioned PBAT product 1,4-butanediol can be synthesized through hydrogenation of succinic acid. Traditionally, succinic acid is derived from transformation of maleic acid in the presence of supported metal catalysts. Biotechnologies now make fermentation of sugars to succinic acid commercial. Therefore, PBAT can be both considered as fossil and bio derived degradable plastic product.

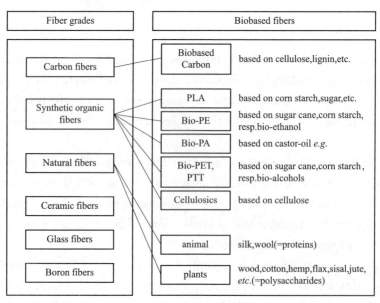

Figure 3-62 Bioplastics and fibers

PLA is produced through ring-opening polymerization of lactide. The starting material for lactic acid can be fermentation products from sugar conversion. Most common starting feedstocks include sugar cane. Technological development is still underway to transform waste lignoncellulosic biomass as well as other low value agricultural wastes. Other lactones and cyclic monomers can be also processed with PLA and similar alphatic esters, which allow good control over thermal, mechanical and chemical properties. Other alternates for renewable plastics also include cellulose-based fibers, succinate polymers. They can be obtained from various other agricultural wastes such as orange peels.

It is important to note that, simply because monomers are produced from renewable materials, does not guarantee improved sustainability. For example, bio-based PET materials can be produced from reaction between ethylene glycol and glucose. However, eutrophication and acidification will give considerable rise in carbon emission and environmental hazards. Bio aromatic diacids still cannot be produced from renewables economically. Other co-polymers derived from sugars display similar properties with PET, but still needs substantial research efforts to address technological challenges.

3.7.3 Using waste chemicals

Reduced carbon emission will be associated with better use of waste molecules from industrial, agricultural and other chemical processes. CO_2 can be incorporated with waste molecules as a co-monomer for producing building blocks for plastics.

Recycling of waste plastics and obtain chemical feedstocks for repolymerization is currently the dominant methodology for upgrading of waste materials. Recycling of PET has been long established at both laboratory and industrial levels. It has been considered as one of the most easiest materials to recycle, second only to aluminum in terms of scrap values for recycling. Methods for PET recycling can be categorized into primary, secondary, tertiary, and quaternary recycling. Primary recycling, known as re-extrusion, deals with scrap materials with similar features to original products. Secondary recycling was commercialized in 1970s. It involves separation of the polymer from contaminants and reprocessing it to granules *via* mechanical force. Tertiary method, more commonly known as chemical recycling, involves the transformation of the PET polymer chain. This method realizes PET reuse through solvolytic chain cleavage back to its monomers or partial depolymerization to oligomers. Water, methanol and ethylene glycol can be used as effective solvents to achieve solvolytic degradation of PET.

Summary

Students should understand, based on the learning of this chapter that, lignocellulose is fundamentally different from traditional fossil feedstocks. High oxygen content and complicated physiochemical properties are key features for bio-derived sources. Basic information on the source of biomass and production status for two important bio-fuels, bio-ethanol and bio-diesel should be fully understood by students as preliminary requirement.

Exercises for Chapter 3

(1) Basic composition of lignoncellulosic feedstocks.

(2) What are the platform molecules for bio-fuels (hints: cellulose, hemicellulose and lignin feedstocks)?

(3) Feedstocks for bio-ethanol and bio-diesel, reaction mechanism and technological bottleneck.

(4) What are the main reactions involved in thermochemical conversion?

(5) Types of lignin and origins. Downstream products from lignin.

(6) Key intermediates for petroleum and bio-based chemicals.

(7) List at least four types of degradable plastic products.

Chapter 04

Fuel Cells: Fundamentals and Technologies

4.1 Introduction

Fuel cells can generate electricity with promising energy efficiency and low environmental impact. Fuel cells are electrochemical devices which convert chemical energy into electrical energy directly. The concept of fuel cells is similar to battery. Batteries are designed as portable device of electrical power and they must carry chemicals necessary to provide the power. Once the chemicals are consumed, batteries can no longer provide energy or power. However, different batteries, a fuel cell is able to provide energy in a continuous mode. Fuel cells do not contain any chemicals, while they simply provide a reaction chamber where fuels and oxidants can react and provide power.

Unlike conventional energy generation devices which produce heat and mechanical work, fuel cells are not limited by thermodynamic limitations of heat engines such as Carnot efficiency. More importantly, since combustion of hydrocarbons in conventional engines is avoided, fuel cells produce sufficient power supply with minimal pollutant.

Fuel cells need to be operated in a continuous mode, bearing resemblance to electrolyzers. Fuel cells can be designed as portable similar to batteries, or for stationary applications for residential and industrial uses.

4.2 History and principles

4.2.1 History

The first set of fuel cell device can be dated back to late 18th and early 19th century. A British scientist Sir Humphry Davy studied and revealed the effect of an electric current on water, to show electrolysis of water is feasible once current is posed for generating H_2 and O_2. In 1802, he showed in a simple device that, O_2 can be added to a chemical cell to generate electric shock. Actually this device was the original form of fuel cell with carbon as anode and nitric acid electrolyte. The German scientist Christian Friedrich Schonbein and British scientist William Grove confirmed that, if water can be split into H_2 and O_2 by introducing electrical power, this process should be reversed and generate electrical energy.

In 1959, another British scientist Francis Bacon studied alkaline fuel cells and were able to develop the cell with 5 kW power for commercial uses. An engineer Henry Ihrig developed a 15 kW fuel cell stack of 1008 cells on alkaline fuel cell to power a tractor. A company later built a golf cart with portable alkaline fuel cells.

NASA invested research efforts in developing alkaline fuel cells for Apollo space program. Fuel cells can supply electrical power and drinking water to astronauts on space flights and space shuttle.

In 1955, two engineers in General electric (GE) developed a different type of fuel cell with proton exchange membrane as the electrolyte. Such solid electrolyte fuel cell eventually replaced alkaline one in late Gemini space program. UK Royal Navy also adopted such unique device for submarine fleet. This was the main driving force to develop fuel cells for automotive power in nowadays.

Energy crisis in late 1970s has stimulated wider research efforts to develop alternative technologies for power generation. New categories of fuel cell devices such as phosphoric acid fuel cells, solid oxide fuel cells and molten carbonate fuel cell have been studied and developed since then. Solid oxide and molten carbonate fuel cells are operated at elevated temperatures to allow fast reaction between H_2 and O_2 molecules. Methanol was also used as substituents for H_2 in fuel cells in 1990s.

The commercialization of fuel cell technologies in civil application has been slow and held back due to high cost of devices. Till 1990s the first commercial stationary fuel cell was launched.

4.2.2 Principles

In a typical fuel cell, two electrodes are spatially separated from each other, with an electrolyte material sandwiched in the middle with close contact with electrodes. Two external gas sources are needed to provide H_2 and O_2 molecules to two electrodes (Figure 4-1). Gas components should diffuse easily through the electrode structures and reach the surface of electrode-electrode interface, where redox reactions occur to generate electric current in external circuit.

Electrolyte in the fuel cell is essentially a semi-permeable **membrane**, which allows a cations or anions to travel through. For example, a proton exchange membrane allows H_2 to pass from one electrode to another but not allowing electrons to diffuse through. But electrons are needed to complete the oxidation of H_2 and reduction of O_2 at interface. Therefore, external circuit is essential for reaction to take place.

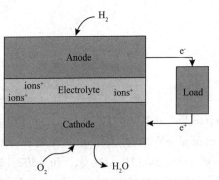

Figure 4-1 Schematic description of fuel cells

In most cases, both fuel and oxidant are gaseous components, thus the electrodes are designed to provide sufficient interface where gas reactants can contact and react to facilitate electrons and charged ions formed during fuel cell operation, followed by diffusion and transport in electrolyte and external circuit.

In H_2 fuel cells, the following reaction occurs and is exothermic. In fact, this is a relatively complex reaction which involves four-electron process and three key steps.

$$2H_2 + O_2 = 2H_2O$$

The first step actually involves oxidation of H_2 molecule. A H_2 molecule is split into two H atoms, each of which releasing an electron to form proton.

$$H_2 = 2H^+ + 2e^-$$

A parallel second partial reaction involves activation of O_2 molecule. O_2 is split into two O atoms, each of which then absorbs two electrons to form O^{2-} species.

$$O_2 + 4e^- = 2O^{2-}$$

However, this reaction cannot occur spontaneously at ambient temperature, as significant energy input is needed to initiate this reaction and cleave O=O bond. In addition, both H—H and O=O bond will dissociate into active atoms at high temperatures thus making this reaction occurring rapidly. To enable redox reaction at relatively low temperature, catalyst is needed to initiate the catalytic circle.

Catalysts are loaded on electrodes. At H_2 electrode (anode), molecular H_2 species are split into H atoms following activation on the surface of catalysts. Subsequently, H atoms are further transformed into H^+ and electron according the equation shown above. H^+ then travel across the electrolyte boundary and reach the surface of O_2 electrode. At O_2 electrode (cathode), O atoms generated from activation of O_2 molecule interact strongly with catalysts. However, reduction of O_2 molecules occur requiring a supply of electrons and H^+ to eventually generate H_2O molecule as the final product. A schematic description of proton exchange membrane fuel cell is shown in Figure 4-2.

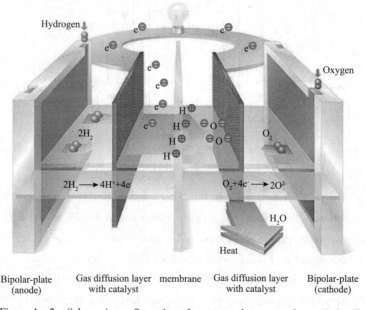

Figure 4-2 Schematic configuration of proton exchange membrane fuel cell

The electrolyte is key for all electrochemical devices. Because electrolytes can filter specific ions from traveling from one electrode to another, thus mixing of cell reactants can be avoided during electrochemical reactions. The electrolyte in a fuel cell is impermeable to gases. And it will not conduct electricity in the form of electron either. In many cases, electrolyte will not conduct negatively charged O^{2-} or OH^- in solid or aqueous phase.

4.3 Fuel cell units

4.3.1 Structure of unit cell

(1) Basic structure

Unit cell is the heart of fuel cell devices. This unit convert the chemical energy into electrochemical energy. The basic structures of a fuel cell device include electrolyte layer in close contact with an anode and a cathode on either side. A schematic figure of a unit cell is presented in the following graph (Figure 4-3) with reactants/products and cation/anion flow direction. As already mentioned in previous section, fuel gases/liquids can be continuously fed to the anode and an oxidant is fed to the cathode in a continuous mode.

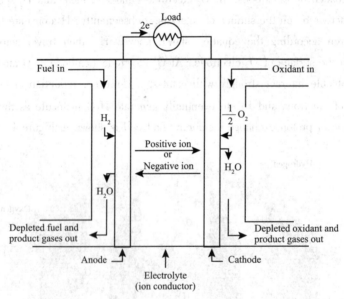

Figure 4-3 Basic structures of acidic fuel cell

The most critical part for unit cells is the three-phase interface on electrode surface, where electrochemical reactions occur with electrolyte, electrons and gas in close contact with each other. The density of these regions and nature of the active sites play the critical role in the overall performances for both liquid and solid electrolyte fuel cells.

For liquid electrolyte fuel cells, reactive gases actually diffuse through the thin film of electrolyte which wets the porous electrode materials and then react electrochemically on the surface of electrode catalysts. Therefore, if the surface of electrode is flooded with excess electrolyte, mass transfer of gaseous reactants might be restrained, poor reaction rate can be observed under such circumstance as a consequence. Therefore, the wetting of electrode surface, contact with electrolyte and mass transfer rate for gaseous components need to be balanced to maintain sufficient electrochemical reaction rates.

In solid electrolyte fuel cells, rational design of large numbers of active sites in close contact with electrolyte and electrode, both physically and ionically is the real challenge. In practical applications, a good interface should consist of well-dispersed active sites of catalyst and selective ion conductivity for H^+ or electrons.

Obviously, the improvement of the three-phase interfaces through sophisticated design of electrode materials including conductivity, porosity, mechanic strength and temperature resistance is key for performance enhancement in commercial fuel cell devices.

In addition, the two electrodes need to be connected with external circuit to ensure electron transport from anode to cathode. Such configuration provides physical compartment to prevent mixing of fuel and oxidant gases.

(2) Additional requirements for electrode materials

To construct cell units, additional requirements and criteria need to be considered:

(a) Transport electrons away from the three-phase interface once oxidation of fuels takes place on electrode surface.

(b) Precise control of feeding rates of fuels and oxidants to ensure stable performances. Both fuels and oxidants should be distributed well in the cell.

(c) Products such as water and other possible by-products need to be separated from bulk phase immediately after reaction occurs.

Therefore, electrode materials should be in porous form to ensure the above-mentioned functionalities. For low temperature fuel cells, noble metal materials are often used to immobilize on the interface of electrode materials for sufficient electrochemical activities. For high temperature fuel cells, bulk electrode materials can be used for stable performances.

(3) Fuel cell stacking

Almost in all practical applications, unit cells need to be combined or stacked in a predetermined fashion to achieve the desired voltage and power output for residential and industrial uses. Stacking of unit cells requires connecting multiple components in series through electronically conductive interconnects.

Planar-bipolar stacking. The most commonly used stack design is the planar-bipolar arrangement. Individual cell unit is electronically connected with interconnects. The basic configuration is plate cells. The interconnects are actually the separator plates with two important functions. One is to provide electrical connection between two adjacent cells and another is to provide gas barrier to separate the fuel and oxidant in adjacent units.

The plate could also be manufactured with channels which can distribute gas flow through the cells (Figure 4-4). The main advantage for planar-bipolar stacking is minimum electric resistance. Depending on the detailed configuration of channels, planar stacking can also be categorized in the followings in terms of gas flow direction.

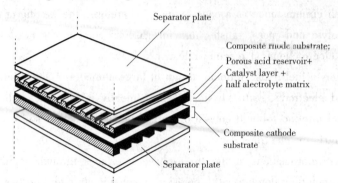

Figure 4-4　Layers for separator plate

(a) Cross flow: fuel and oxidant flow perpendicular to each other.

(b) Co-current flow: fuel and oxidant can flow in the same direction.

(c) Counter-current flow: fuel and oxidant flow opposite direction.

(d) Serpentine flow: fuel and oxidant flow in zig-zag direction.

(e) Spiral flow: for circular cells.

In all, the bipolar stacking can be achieved through internal, integrated and external manifolds to combine unit cells.

Tubular cell stacking. This type of stacking is useful for high temperature fuel cells. Tubular cells show advantages in sealing and structural integrity compared with planar stacking. But the unique geometric challenges to stack designer are achieving high energy density and short current path. To minimize the length of electronic conduction for each cell, sequential series connected cells have been developed.

(4) Fuel cell as a whole system

In addition to the mode of stacking for unit cells, other components also require special attention to achieve stable and secure power output. Fuels should be pretreated except for pure H_2. Fuel preparation is needed to remove impurities and thermal conditioning. Sometimes, fuel processing is needed. For example, a reformer can be installed to enable fuel and oxidant reaction and generate H_2-rich anode feeds. The quality and stable supply of air is also important to maintain certain reaction rate and power output. Air filters are often required to control oxidant quality. Electrochemical reaction of fuel and oxidant releases significant amounts of heat, especially under high current density. Therefore, thermal management is also critical for durable operation of fuel cell devices to maintain suitable temperature.

Water is a clean by-product but water management is very important for fuel cells. To ensure stable and smooth operation, water removal should be carefully considered for stacking units. In addition, the power conditioning equipment is needed to provide variable voltage output which can be directly used for external load.

The following figure illustrate a typical fuel cell device with natural gas as feedstock to provide H_2-rich fuel for power generation unit (Figure 4-5). The conventional fuel needs to be cleaned prior to reforming reaction for H_2 generation. Air is blowing into the stacked cell units with predetermined rates. The power output is largely dependent on the number of cell stacked together. The power conditioning unit converts the electric power from DC to AC for residential or industrial uses.

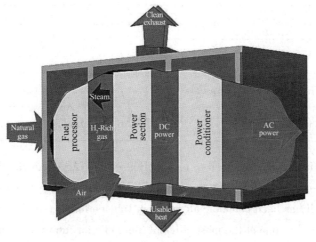

Figure 4 – 5 Typical power generation unit

4.3.2 Types of fuel cell: general description

A variety of fuel cells have been developed in the past few decades. General classification of fuel cells is based on the type of electrolyte used in units (Table 4 – 1). (a) **polymer electrolyte fuel cell** (PEFC), (b) **alkaline fuel cell** (AFC), (c) **phosphoric acid cell** (PAFC), (d) **molten carbonate fuel cell** (MCFC) and (e) **solid oxide fuel cell** (SOFC). The types of electrolyte actually determine the operating temperatures of fuel cells. For aqueous phase electrolyte, operation temperature should not be higher than 200°C due to serious degradation of electrolyte and high vapor pressure imposed on the system. For low temperature devices, H_2 needs to be purified prior to entrance of the unit cell. More importantly, expensive catalysts such as Pt materials need to be rationally designed as they can be easily poisoned by traces CO. In high temperature fuel cells, CO and CH_4 can be internally converted into H_2 species or directly oxidized in unit cells to generate electric power. The following table summarizes the main differences for various fuel cells.

Table 4 – 1 Comparison of various fuel cells

	PEFC	AFC	PAFC	MCFC	SOFC
Electrolyte	Hydrated polymeric ion exchange membrane	Mobilized immobilized potassium hydroxide in asbestos matrix	Immobilized phosphoric acid in SiC	Immobilized molten carbonate in $LiAlO_2$	Perovskites ceramics
Temperature/°C	40 ~ 80	65 ~ 200	200 +	650 +	600 ~ 1000
Power/kW	1 ~ 100	1 ~ 100	150 ~ 200	600 ~ 700	500 ~ 1000
Efficiency/%	40 ~ 60	>60	40	50	60
Electrode	Carbon	Transition metals	Carbon	NiO_x	Perovskite

Continued

	PEFC	AFC	PAFC	MCFC	SOFC
Catalyst	Pt	Pt	Pt	Electrode materials	Electrode materials
Interconnect	Carbon or metal	Metal	Graphite	Stainless steel or Ni	Ni, ceramic or steel
Charge carrier	H^+	OH^-	H^+	CO_3^{2-}	O^{2-}
Water management	Evaporation	Evaporation	Evaporation	Gas	Gas
Thermal management	Gas-liquid cooling	Gas-electrolyte circulation	Gas-liquid cooling	Internal reforming	Internal reforming

PFEC was the first type of fuel cell to be widely used in NASA launch missions. The cell can be operated under relatively milder temperature and is very efficient if provided with high purity H_2 fuel. This type of cell has been commercially used in transportation vehicles due to its low weight.

AFC is considered as one of the most efficient fuel cells for future implementation. It operates under low temperature. It has been often used in U.S. space program. However, such fuel cell is relatively more expensive due to use of Pt noble catalysts. Therefore, most recent efforts have been dedicated to make more cost-effective Pt catalysts such as PtCu, PtCo, PtNi, *etc*.

PAFC has been commercialized early in 1990s. But this technology is more expensive than others. MCFC is operated at high temperature which converts natural gas reforming. Such fuel cell has more complicated internal structures thus only utilized for larger size with high power output. It is the most successful application for stationary units. SOFC is another type of high temperature fuel cell. The unique solid state electrolyte makes the cell unit relatively robust compared with other types. This fuel cell can be manufactured for both small and large scale devices.

Methanol fuel cells have also been used as methanol rather than H_2 can be directly burned for power generation. This type of fuel cell is useful for portable devices such as mobile phones, computers. Other alcohols such as ethanol, propanol and ethylene glycol can also be used as fuels in PEFC type of fuel cell.

(1) PEFC

An **ion exchange membrane** (fluorinated sulfonic acid polymer) is used in PEFC with excellent proton conductor. The only liquid component in this type of cell is water. Therefore, corrosion is insignificant. In most cases, carbon electrodes immobilized with Pt metal are used as catalyst materials for both anode and cathode.

But water management is a critical issue in PEFC, because it greatly affects the performances. The cell unit needs to be operated at temperature not higher than the evaporation of water. In other words, the evaporation rate should not be higher than the formation of water as the product. This is because the membrane needs to be hydrated for H^+ conduction. Also, the polymer membranes cannot operated under harsh temperature, typically around 60~100℃.

Moreover, high purity H_2 with minimal or no CO is used thus extensive fuel pretreatment is need to satisfy this requirement. Additionally, high Pt loading as high as 30% is needed for both anode and cathode.

The main application of PEFC lies in automobiles and portable devices in the past decade, although many attentions have also been drawn for stationary applications. Numerous efforts have been put to pursue prime power for vehicles. The investment of PEFC has surpassed all other types of fuel cells.

Advantages: Solid electrolyte avoids use of liquid materials and provides excellent resistance to gas crossover. Low operating temperature also allows quick start-up for engines and alleviate corrosion issues for acidic medium. Test results have shown that the power density for PEFC can be 2 kW/L.

Disadvantages: Water management is a serious issue for PEFC. Sufficient membrane hydration needs to be achieved to balance against flooding the electrolyte. It is difficult to use waste thermal energy as the operating temperature is very low and narrow for PEFC. Impurity in H_2 feedstocks is another significant issue. CO and sulfur species should be removed completely. Therefore, fuel pretreatment is important for PEFC, which cause unfavorably high cost for gas cleaning and complication of whole unit cells. Use of Pt noble metal as catalysts also increase the overall cost of PEFC.

(2) AFC

Typically, highly concentrated KOH is used in AFC with weight percentage up to 85%. The operating temperature can be as high as 250℃. For operating temperature lower than 140℃, ~ 50% of KOH aqueous solution can be used. The electrolyte is retained in asbestos matrix. Various materials can be used such as Ni, Ag, metal oxide can be applied as electrocatalysts. CO_2 will react with alkaline electrolyte and form carbonate salts, altering the nature of electrolyte. Therefore, the fuel should be processed to remove acidic impurities before entering the cell unit.

AFC has been very successful in space program. However, its larger application in industry has been hampered by high sensitivity to CO_2 species in fuels. Companies in Europe and US are still interested in developing more advanced AFC technology for portable devices.

Advantages: Compared with other fuel cells, the kinetics for cathode reactions, namely O_2 reduction reaction, is better, thus displaying great promises for finding more cost-effective catalyst materials.

Disadvantages: The high sensitivity to CO_2 for electrolyte is the main disadvantage for AFC. Therefore, both fuel and O_2 (air) supplies need to remove CO_2 and CO components. This unfavorably increase the size and cost for AFC.

(3) PAFC

Highly concentrated phosphoric acid can be used as electrolyte. PAFC is typically operated at 150~220℃. At relatively lower temperature, the ion conductivity for phosphoric acid is not good and CO poisoning of Pt catalysts remains a big issue. Stability for phosphoric acid is relatively higher than other types of acid. More importantly, use of 100% of phosphoric acid also avoid the use of water and minimize evaporation pressure. Thus water management is not difficult in PAFC. The commonly used matrix is silicon carbide, with Pt catalysts for both anode and cathode.

PAFC has been mostly used for stationary applications. Numerous test demonstrations have been very successful in Japan and US, but the development for PAFC is still very slow.

Advantages: Tolerance for CO is very high for PAFC, up to 1%. In addition, the operating temperature is low enough for many common construction materials to sustain long-term uses. It provides considerable convenience for thermal management. Typically, the efficiency for PAFC is around 37%~42%, which is higher than most PEFC systems. The heat generated from PAFC can also be directly reused for residential and industrial purposes.

Disadvantages: Kinetics for cathode side reaction is very poor compared with other fuel cells. Expensive Pt catalysts are mostly needed. Fuel processing is still required to remove water content in feeds. Relatively high temperature and corrosive nature of phosphoric acid demand special attention and investment for construction materials.

(4) MCFC

For MCFC, the electrolyte is often alkaline carbonate adsorbed on $LiAlO_2$ matrix. The typical operating temperature is 600~700℃, which is significantly higher than PEFC, AFC and PAFC devices. Under such condition, alkaline carbonates form conductive molten salts, with carbonate ion providing ion conduction in the cell. Ni (anode) and nickel oxide (cathode) are active to promote the redox reaction. Noble metal catalysts are not necessary in MCFC. In addition, many types of hydrocarbons can be processed in this type of cell.

The main end applications for MCFC are stationary units and marine purposes. Large size and slow start up are not considered as serious issues for MCFC. Both fossil and renewable fuels can be processed in MCFC, with extensive demonstrations for stationary applications. Compared with PAFC, MCFC display more extensive potentials for large scale applications in stationary devices.

Advantages: harsh operating temperatures for MCFC actually results in several benefits, including inexpensive Ni catalysts to promote redox reactions, CO and hydrocarbons are both acceptable for MCFC, because they can internally be transformed into H_2 via reforming and gas shift reactions. Thus the overall efficiency could be around 40%~50%. High temperature also enable good use of heat generated from the cell, which further boost the energy efficiency up to 60%.

Disadvantages: The main challenge for MCFC applications is use of corrosion and temperature resistant materials. High-grade stainless steel is needed for MCFC and only inexpensive metals could be applied under such harsh temperature. In addition, high temperature operation could also cause mechanical issues and poor stack life. High contact resistance limits the power output to around 100~200 mW/cm^2 at practical scale.

(5) SOFC

Y_2O_3-ZrO_2 based materials with porous structures are often used as the solid electrolyte for SOFC. The cell is typically operated at 600~1000℃ where O^{2-} is the conductive ion. Anode materials are Co-ZrO_2, Ni-ZrO_2 and other non-noble metal species, while cathode materials are Sr-$LaMnO_3$ catalysts.

Early work has been focused on studying SOFC operated around 1000℃. Recent attention has been paid to developing thin-electrolyte cell devices with improved cathodes to reduce the temperature to around 650~850℃, which enable fast development of low-cost construction materials for high-performance SOFC. SOFC has been widely used for stationary applications, mobile power and auxiliary power for vehicles.

Advantages: Since the electrolyte is in solid phase, the cell can be cast into a variety of different shapes such as tubular, planar and monolithic. Ceramics are often used to alleviate the corrosion problems in SOFC. Interfacial engineering to ensure three-phase boundaries at electrolyte-fuel-electrode and avoid sliding movement of electrolyte is critical for stable operation under high temperatures. No requirement for CO_2 processing is needed in SOFC. Thin-electrolyte SOFC have been found to show similar energy density as with PEFC devices. In addition, high operating temperatures also favor heat reuse thus the overall energy efficiency is also around 60% for SOFC. In addition, fast kinetics avoids the use of expensive metal catalysts.

Disadvantages: Thermal expansion at high temperature causes structural mismatch in construction materials, thus leading to possible fouling of devices. Therefore, sealing between cells is very difficult in planar configurations. Obviously, harsh operating temperatures poses challenges in fabricating and selection of suitable structural materials for SOFC. And corrosion of metal stack components is another issue. As a result, the life of stack materials and catalysts also affect the overall cost of SOFC.

Summary

According to the performance analysis for various fuel cells, it is found that electric power generation combined with lower heat reuse is necessary to improve overall efficiency. Therefore, hybrid fuel cells/turbine cycles providing efficiencies higher than 60% have been demonstrated to be applicable for commercial purposes. Fuel emission is one of the important parameters to evaluate the efficiency of fuel cells (Table 4-2). Another important feature of fuel cells is that, the cost and performances are less dependent on the size or the scale of devices, which is different from other power generation technologies. This unique features opens up new applications for fuel cells to install where conventional power generation technologies are not practical and economical. Other characteristics of fuel cells also include, direct energy conversion (no combustion), no moving part in generator, quiet, relatively low temperature, siting ability, fuel flexible, remote operation, size flexibility and rapid load capability. The unfavorable negative features for fuel cells include high market price, not durable compared with conventional power generation units, no large-scale infrastructure.

Table 4-2 Summary of main fuel constituent impact on various fuel cells

Gas	PEFC	AFC	PAFC	MCFC	SOFC
H_2	Fuel	Fuel	Fuel	Fuel	Fuel
CO	Poison	Poison	Poison	Fuel	Fuel
CH_4	Diluent	Poison	Diluent	Diluent	Fuel
CO_2 and H_2O	Diluent	Poison	Diluent	Diluent	Diluent
H_2S or COS	No data	Poison	Poison	Poison	Poison

4.3.3 Applications

The main applications for fuel cells are stationary electric power plants, motive power devices, on-board electric power generators. Selected applications are discussed below.

(1) Stationary electric power

As mentioned earlier, the energy efficiency for fuel cells is almost independent on the size of the units. Therefore, initial development of stationary plants has been focused on several hundred kW to low MW capacity. Small devices such as kW to MW can also be sited adjacent to users' facility thus suitable for cogeneration operation. Larger units of 10 MW devices are more often used for distributed systems. Such plants are often fueled with natural gas or shale gas. Once the plants are implemented and commercialized leading to price reduction and improvement of efficiency, fuel cells are regarded as most promising power generation technology as the efficiency is significantly higher than conventional electric power generation. The fuel product from as coal gasifier, once cleaned, is compatible for use with fuel cells.

In general, the on-site plant is more efficient in providing economic and beneficial electric power input for operating facilities such as commercial buildings and industrial plants, also for hospitals, hotels, large office buildings, manufacturing plants, waste water treatment plants and schools/institutes. Overall the following features are listed as key characteristics: on-site energy, continuous power, uninterrupted fuel/power supply and independent lower source.

US Department of Energy has published one handbook in 2004, illustrating the typical features of various commercialized stationary devices and their corresponding features to meet the basic requirement for residential and industrial uses.

(2) Distributed generation

Distributed systems usually involve small modular power generation devices, which are sited close to downstream facilities. The typical power is lower than 30 MW with use of clean fuels. For example, gas turbines, biomass-based power generators, solar panels, photovoltaic systems, fuel cells, *etc*. The size and efficiency for typical distributed generators using fuel cell technologies are summarized in the Table 4-3.

Table 4-3 Comparison of size and efficiency of various fuel cells

Type	Size	Efficiency/%
Reciprocating engines	~6 MW	30~37
Micro turbines	10~300 kW	20~30
PAFC	50 kW~1 MV	40
SOFC	5 kW~3 MW	45~65
PEFC	~1 MW	~40
Photovoltaics	1 kW~1 MW	-
Wind turbines	150 kW~500 kW	-
Hybrid renewables	~1 MW	~50

Due to the pandemic of COVID-19, there is a strong growth in the market for distributed generation systems, which is aimed at customers highly relying on durable energy supplies, including hospitals, plants, schools, groceries, restaurants and bank facilities. It is estimated that the growth is 5 ~ 6 GW per year in North America, even larger number for Asian and EU areas.

Among many applications of distributed systems, the following ones are particularly important.

(a) Peak shaving. The cost of power generation fluctuates dramatically depending on the hour during the day and night owing to varying demand and generation. Therefore, customers could use distributed systems during high-cost and peak hours.

(b) Cogeneration. The heat cogenerated by fuel cells are useful for additional heat requirement in remove areas, particularly for sites with all day demands for thermal/electric applications.

(c) Grid support. Strategic implement of distributed devices can provide for reducing the overall cost for upgrading existing electric systems and supplying more energy demanded in regions where new power needs are generated occasionally.

(d) Standby power. Reliable energy back up is essential for customers. Power generation is particularly important during power outrage. Electric power can be provided until regular service is restored.

(e) Remote devices. The flexibility of distributed devices is important for remove applications for mobiles and portable units.

However, there are still several issues plaguing the larger applications of distributed devices. For example, technical barriers, economic and regulatory issues are main problems. Questions are been raised for reliable connection with existing grids. Lack of standardized procedures for implement and uses has created unfavorable market delays and discouraged customers from investing the relevant projects. Considering the current status of applications, PAFC and SOFC have been successfully commercialized, while PEFC is still under research development for distributed systems.

(3) Motive power devices

There has been a super strong push to develop fuel cell vehicles for light-duty and heavy-duty uses since late 1980s. The main driving force for such development is the need to clean and sustainable transportation systems. H_2 has the advantage of clean burning and no emission to the environment. Alternative fuels, such as methanol, ethanol, propanols and natural gases have been applied for fuel cells in cars, trucks and public transient vehicles of everyday uses. Another advantage include the low maintenance of fuel cell vehicles owing to fewer moving parts. As a result, the development has been sponsored by major automotive companies in North America, EU and Japan.

Toyota and Honda have successfully commercialized fuel cell vehicles after late 2000s. Other major automobile manufactures including General Motors, Volkswagen, Volvo, Chrysler, Nissan and Ford have also commercialized PEFC vehicles operating on H_2, methanol and gasoline.

(4) Other applications

The application of AFC has been demonstrated for space programs. AFC devices with 1.5 kW to 30 kW energy output have been widely used in various NASA launch missions. H_2/O_2 is the main fuel and oxidant combination in space programs, while methanol fueled cells with 80 kW have been applied in submarine for portable power systems.

Auxiliary power units are also suitable for large-market uses. Because such units are flexible and do not require high-demand and strict specification for typical stationary devices and grid-incorporated distributed systems.

Overall, SOFC and PEFC have been established for demonstration status for refined and commercial prototype of plants. PEFC has been more extensively implemented for residential uses rather than auxiliary uses.

4.4 Fundamentals for fuel cells

4.4.1 Modern electrochemistry

(1) Fundamentals

The maxim electric power output (W_e) obtained from one fuel cell is represented by the following equation. The work is dependent on Gibbs free energy (ΔG), number of electrons involved in the reaction (n). F is Faraday's contant (96487 C/mol e$^-$) and E is the theoretical potential of the cell.

$$W_e = \Delta G = -nFE$$

The Gibbs free energy is also given with enthalpy change (ΔH) and entropy change (ΔS). Clearly the available energy for is limited by $T\Delta S$, which represents the unavailable energy resulting from entropy change.

For a general cell reaction $\alpha A + \beta B \rightarrow \gamma C + \delta D$, the standard state Gibbs free energy change of the reaction is $\Delta G^\circ = \gamma G_C^\circ + \delta G_D^\circ - \alpha G_A^\circ - \beta G_B^\circ$.

The Gibbs free energy change of the reaction can be expressed by the following equation:

$$\Delta G = \Delta G^\circ + RT\ln\left(\frac{f_C^c f_D^d}{f_A^a f_B^b}\right)$$

In this equation, f represents the fugacity of certain species. By substituting the original Nernst equation, we obtain the following equation:

$$E = E^\circ + \frac{RT}{nF}\ln\left(\frac{f_A^a f_B^b}{f_C^c f_D^d}\right)$$

The output potential E is dependent on the operating pressure and temperature. The value of E is also dependent on the fugacity of products and reactants.

Nernst equation gives the maxim open circuit potential E for the fuel cell. For various types of fuel cells, different anode and cathode reactions are involved, leading to varied cell potentials during operation.

Actually Nernst equations provides the relationship between ideal standard potential E° and the ideal equilibrium (open circuit) potential E considering the fugacity of reactants and products.

For overall cell reaction, the cell potential increases with partial pressure of products. For example, H_2 fuel cells, the ideal cell potential is enhanced under higher reaction temperature and pressure. Thus performances of fuel cells can be improved. If carbon-containing fuel is used in the anode reaction, CO_2 produced will also participate the cathode reaction to maintain invariant carbonate concentration in the electrolyte. However, the concentrations in the anode and cathode are not necessarily identical, in MCFC, for example, CO_2 partial pressures for both electrodes need to be present in Nernst equation.

The ideal standard cell pressure is 1.23 V for H_2/O_2 fuel cell at 25℃. The value is 1.18 V for gaseous water product. This is because there is difference of Gibbs energy for vaporized water at standard condition. The following graph (Figure 4 – 6) illustrates the temperature dependence of cell potential.

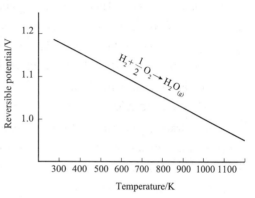

Figure 4 – 6　Reversible potential of H_2 oxidation at different temperatures

The operating voltage is strongly dependent on the concentration of reactants. The maxim voltage occurs when both anode and cathode are fed with pure fuels and oxidants, respectively. Concentration gradient between entrance and exit also reduce the operating voltage of fuel cells. The decrease in partial pressure also lead to a 0.25 V reduction for operating cells.

When CO and CH_4 are used as fuel, H_2 is produced through reforming and water gas shift reactions. Therefore, in the actual cells, more than one type reactions will occur to ensure more H_2 is produced for anode reaction. Thus the operating conditions strongly affect the overall efficiency.

(a) The open cell potential for oxidation of CO and CH_4 is lower that of H_2 cells.

(b) More surface active sites are available for reforming and water gas shift reactions on the surface of anode materials, compared with the three-phase boundary for electro-oxidation.

(c) Mass transfer rate of hydrocarbons are much slower compared with H_2, leading to more significant concentration polarization under high electric current conditions.

The direct oxidation reaction is important for improving overall performances of fuel cells. This is because the degree of oxidation affects the reaction temperature thus influencing the extent of reforming and shift reactions. As a result, the plant complexity and cost will be estimated to adjust the target performance of fuel cells. Some state-of-art units consist of external reformers to ensure more H_2 is produced from CO. In some internal reforming cells, active sites and oxidative anode are close to each other to promote the tandem reactions.

(2) Kinetics for electrochemistry

In general, the kinetics for a certain chemical reaction can be described using Arrhenius equation,

$$k = k_o e^{-\frac{E_a}{RT}}$$

where k is the rate constant and k_o is the pre-factor. Based on the general definition, the output potential for fuel cells actually determined by reduction and oxidation rates, namely as O_2 reduction and H_2 oxidation. The equation for output potential can be modified as follow:

$$E = E^o + \frac{RT}{nF}\ln\left(\frac{k_{oxidation}}{k_{reduction}}\right)$$

Where the K is actually defined as $\frac{k_{oxidation}}{k_{reduction}}$. The rate constant at cathode and anode are not equally in most cases. Therefore, the potential is determined by the equilibrium of the redox reactions. Different from thermocatalysis, energy in electrocatalysis systems can be changed by applying varied voltage to overcome kinetic limitations. As a classic example, water splitting to H_2 and O_2 can be achieved at 1.48 V at room temperature. But in reality 1.7 V is needed to overcome the activation barriers.

(3) Mass transfer

Mass transfer in electrochemical reaction systems often follow three routes, diffusion, convection and migration. Diffusion in electrochemical medium is determined by Fick's law. The diffusion flux J occur from high to low concentration in a homogeneous solution. The flux is proportional to the concentration gradient of species A along z direction.

$$J_o = -D\frac{\partial C_A}{\partial z}$$

Simple diffusion parameters can be measured with cyclic voltammetry and applied with Butler-Volmer equation. Fick's second law also involves time dependent behaviors, which can be used to model diffusion rates.

Convection is enhanced by bulk motion of fluid. Such motion of bulk fluid is enforced by pressure gradient in electrochemical systems. In most cases, the convection mass flow is in laminar flow and easily predicted. Navier-Stokes equations are often used to solve the convection flow rates. As the third source, molecular migration in electric field also contributes to mass flow. The voltage gradient across the cell is a potent force driving the charged ions from one electrode to another.

4.4.2 Charged electrode interface

The surface chemistry of electronic layers on electrode has been well illustrated in Chapter 2. The double layer at the interface of solid electrode and liquid medium can consist of a positively charged one on the surface of negatively charged electrode, and a negatively charged layer in solution.

4.4.3 Cell efficiency

The thermal efficiency of a fuel device is defined as the useful energy (work) produced during the enthalpy change.

$$\eta = \frac{work}{\Delta H}$$

In conventional fuel converting units, fuels are transformed into heat, which is then converted into mechanical energy and electric energy. For a thermal engine, Carnot showed that the maximum efficiency is limited by the ratio of absolute temperature at which heat is rejected and absorbed.

Fuel cells convert chemical energy directly to electric power. In the ideal fuel cell, The change of Gibbs energy represent the maximum useful energy available at the temperature of reaction. The efficiency is thus defined as:

$$\eta_{ideal} = \frac{\Delta G}{\Delta H}$$

For H_2 fuel cell, the change for Gibbs energy at 25℃ is approximately 237.1 kJ/mole while the thermal energy (ΔH) is 285.8 kJ/mole. Therefore the ideal efficiency is estimated as follow:

$$\eta_{ideal} = \frac{\Delta G}{\Delta H} = \frac{237.1}{285.8} = 0.83$$

However, in carbon fuel cells, the value for Gibbs energy change is higher than enthalpy change, thus alternative definition is proposed as the ratio of operating potential to the ideal cell voltage. For example, the thermal efficiency for a H_2/O_2 fuel cell is expressed as:

$$\eta = \frac{work}{\Delta H} = \frac{work}{\Delta G/0.83} = \frac{V \times current}{E_{ideal} \times current/0.83} = \frac{0.83V}{E_{ideal}}$$

Since the ideal potential of fuel cell at 25℃ is 1.229 V, the efficiency of actual cell is given as 0.675V.

If the fuel is not completely oxidized at anode, the voltage efficiency will be lower than the above-mentioned equation. For example, SOFC operated at 800℃ can achieve 54% cell efficiency when fuel utilization is 90% (Figure 4-7).

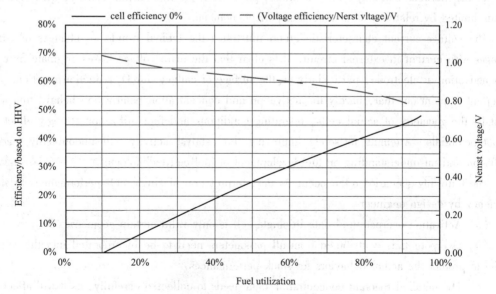

Figure 4-7 Efficiency loss with fuel utilization

The actual cell potential is decreased from it ideal potential because of several irreversible losses (Figure 4-8). The losses are often referred as **polarization**, overpotential, *etc*. Ohmic losses are also important for voltage decrease. The following three phenomena contribute to the actual power losses in fuel cells.

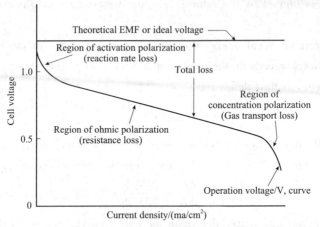

Figure 4-8 Voltage loss with fuel utilization

(a) **Activation overpotential**. The sluggish intrinsic kinetics at anodes and cathodes causes losses in actual cell potential. Efforts are needed to improve the reaction rates by modifying the surface and structural properties.

(b) **Ohmic losses**. The ionic resistance in the electrolyte and electrodes results in voltage losses. Interconnects, current collectors and other contact parts have such effect on the overall voltage of actual cells.

(c) Mass transport losses (**concentration overpotential**). It is caused by the finite mass transfer rate between electrolyte and electrode, particularly on high current density, where reaction rate is limited by relatively slow mass transport phenomena.

The voltage-current characteristic graph illustrates the typical trends for changes of output voltages with current in external circuit. It is clear that the main voltage losses originate from the poor activation at electrode, particularly on the surface of cathode for O_2 reduction reaction. It is important to point out that, hardly the activation and concentration polarization behave this way. Actually the geometry of actual cells, operating conditions and fuel utilization strongly affect the output potentials. Attempts to solely improving the catalyst activity will undoubtedly lead to confusion and misunderstanding on the development of fuel cell technologies.

The following scenarios often occur as misinterpretation of single cell performances to stack efficiency by design engineers.

(a) Activation overpotential data in single cell is only valid certain geometries.

(b) Overpotentials contributed from all parameters need to be incorporated into the refined model to predict the actual behaviors for stack performances.

(c) The detailed reactant concentration data needs to collected carefully, as it will affect the interpretation of activation loss information as well as the overall efficiency of cells. However, it is only possible to acquire the data under low fuel utilization conditions.

(d) Analysis of single cell performances is more reliable when different cells are compared to exclude the influence of interconnects, variation and fluctuation of operating conditions.

The physical and description of voltage losses involved in fuel cell operation actually leads to the derivation and development of several types of models to represent the behaviors in the overpotential curves. Since many steps and processes are included in fuel cell operation, obtaining sufficient experimental data is the priority. In addition, each step, such as surface reaction, diffusion and ionic conduction requires derivation and solution of differential equations. However, coupling both transport and chemical reactions in cell model can be very challenging. Therefore, the advanced description of the following points should be considered first.

(a) Gas diffusion towards electrodes.

(b) Electrochemical reactions on the surface of electrodes.

(c) Transport of protons from anode to cathode.

(d) Formation and drag of water molecules due to the effect of solvation shell around protons.

(e) Mixing and back diffusion of water molecules due to difference of vapor pressure between anode and cathode.

(f) Transportation of water molecules in both liquid and gaseous medium at cathode and anode surface.

(g) Gas crossover of H_2 from anode to cathode.

The material and energy balances in fuel cell units are basis for model development. A schematic description of the above processes is presented in Figure 4–9.

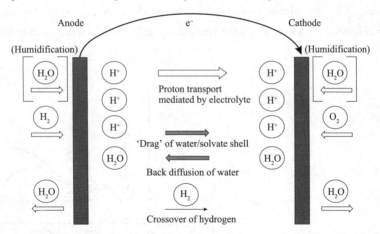

Figure 4–9 Materials and energy balance in fuel cells

(1) Activation overpotentials

It is caused by the poor intrinsic kinetics on the surface of electrodes. The electro chemical reactions involve transfer of several electrons and each of the elementary step demand certain energy to overcome the activation barriers. As already mentioned, three factors all contribute to the voltage losses, but in the case of electrochemical reaction occurring with activation polarization as high as 0.05~0.10 V, it is possible to assign the voltage losses to activation overpotential. To approximate such voltage drop, a semi-empirical equation, Tafel equation, can be applied:

$$\eta_{act} = \frac{RT}{\alpha nF} \ln\left(\frac{j}{j_o}\right)$$

Where α is the electron transfer coefficient of the reaction at the electrode, and j_o is the exchange current density. The Tafel plots provide a visual understanding of the activation polarization of the catalyst in the fuel cell. The extrapolated intercept at zero polarization is the measure of maximum current which can be derived under negligible polarization. The slope of the curve can give the transfer coefficient (Figure 4-10).

(2) Ohmic overpotentials

Ohmic losses occur due to the ionic resistance of electron flow in the electrolyte and electrode. The ohmic losses can be quantitatively expressed by the following law:

$$\eta_{ohm} = iR$$

Where i is the current through the cell and R is the total cell resistance, which obviously include electronic, ionic and contact resistance. In different types of fuel cells, the contribution from the three types of resistance varies significantly. For example, in planar stacks, ionic resistance dominates, while in tubular structures bulk resistance is often significant.

Experimental studies can employ various methods to quantify the contribution of ohomic resistance in fuel cells. Looking at V-I curve reveals the linear section of performance characteristics (Figure 4-11). The linear portion of the curve often reflects the value of specific resistance. A more accurate method can be applied to determine ohmic resistance, namely impedance spectroscopy. The ohmic resistance is the real value of the impedance of the point where imaginary impedance is zero. Clearly, ohmic resistance is independent on gas concentration in fuel cells. However, mass transfer and kinetics change remarkably with varied feed concentration at anode.

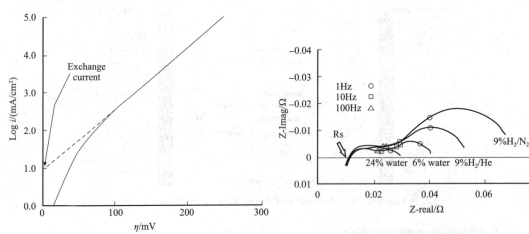

Figure 4-10 Activation polarization   Figure 4-11 Impedance spectrum of anode in SOFC

Given the geometry parameters of fuel cells, and the properties of the material used to construct the device, one can measure and calculate the resistance of the whole device. The data from measurement and calculation can be compared to ensure sufficient accuracy to determine the resistance of stack structures. The contact resistance cannot be evaluated from the fundamental data. It is often assessed by the difference in measured and calculated data.

(3) Concentration overpotentials

Fuel oxidation and O_2 reduction reactions are catalyzed by solid catalytic materials. Therefore, gas-liquid-solid interfacial mass transfer is involved in such systems. The finite mass transportation rate will affect the overall reaction efficiency as reactants will be consumed on the surface of electrodes. As a result, the concentration gradient exists at three-way phases of electrode materials. In SOFC, gas diffusion determines the actual reaction rates. In PEFC, diffusion of water in liquid phase affect the actual rates of O_2 reduction.

Under practical conditions, high density and low fuel and O_2 concentration actually lead to significant loss of output potential due to mass transfer overpotentials. Fick's law can be used to describe diffusion phenomena:

$$i = nFD \frac{dC}{dz}$$

It is clear that the current is limited by the diffusion rate, which is determined by D (diffusion coefficient) and concentration gradient. If the bulk and surface concentration of reactant are C_A and zero respectively, then one obtained the theoretic maximum mass transfer rate and limiting current ($i_{\lim}$) as a result.

$$i_{\lim} = nFD \frac{C_A}{\delta}$$

At equilibrium condition, where no current is flowing the potential can be expressed as:

$$E = E^\circ + \frac{RT}{nF}\ln(C_A)$$

If there is a current flowing and surface concentration of is lower than bulk concentration, the Nernst equation is changed into:

$$E = E^\circ + \frac{RT}{nF}\ln(C_{A,\text{surface}})$$

Therefore, the difference produced by concentration gradient is defined as concentration polarization:

$$\Delta E = \eta_{\text{con}} = \frac{RT}{nF}\ln\left(\frac{C_{A,\text{surface}}}{C_A}\right) = \frac{RT}{nF}\ln\left(1 - \frac{i}{i_{\lim}}\right)$$

In most cases, when high charge transfer rate is achieved, concentration polarization become dominant, where the contribution from activation polarization is negligible.

(4) Cumulative effect

The combined effect from activation polarization, ohomic resistance and concentration polarization cause losses of voltage as the following equation:

$$\eta = \eta_{\text{act}} + \eta_{\text{ohm}} + \eta_{\text{con}}$$

Due to the polarization, the output potentials for anode and cathode become:

$$V_{\text{anode}} = E_{\text{anode}} + \eta_{\text{anode}}$$
$$V_{\text{cathode}} = E_{\text{cathode}} - \eta_{\text{cathode}}$$

As a result, the output voltage is summarized as:

$$V_{\text{cell}} = V_{\text{cathode}} - V_{\text{anode}} - iR - \eta_{\text{con}}$$

The goal of fuel cell design is to minimizing the significance of polarization through modification of cell structures, materials, feedstock properties, *etc*. For all cells, compromises exist between achieving high performances by operating under harsh temperature and pressure, and issues brought by stability and durability of cell components.

4.4.4 Mechanism for voltage losses

Three types of overpentials can be quantitative assessed using model equations. However, in-depth understanding on molecular level of loss mechanism is essential to develop reliable models to accurately describe the behaviors of polarization. Critical parameters of derived models will be corrected for improved prediction.

(1) Losses at open circuit

In most fuel cells, the actual voltage is often 200~300 eV lower than the Nernst voltage. The losses in voltage can be assigned to the following reasons:

(a) Gas crossover between anode and cathode, leading to a mixed potential.

(b) Inconsistent catalyst composition.

(c) Gas solubility on the surface of catalysts in the liquid film.

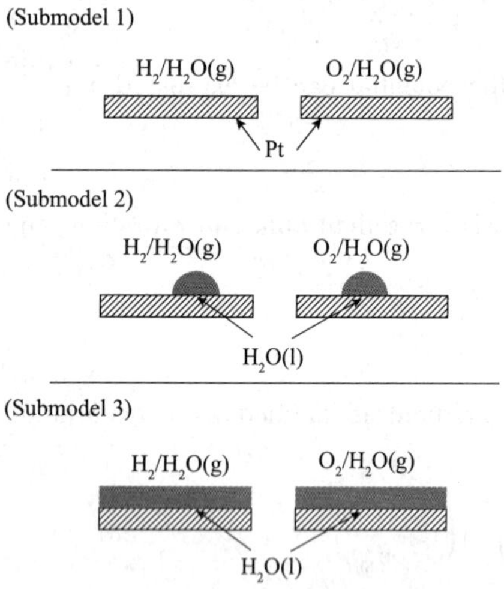

Figure 4-12 Proposed models for open circuit loss

The impact of open circuit on the actual voltage losses is difficult to be quantified, depending on the fundamental understanding on the behaviors at open circuit. In this part, three different submodels are proposed to illustrate such losses (Figure 4-12).

Submodel 1 considers that the surface of catalyst is mainly covered by gaseous component. Therefore, only gas composition is needed to calculate the open circuit losses. Submodel 2 assumes that, there exist gas-liquid equilibrium on the surface of electrode, where catalyst is partially wet by liquid water. As a result, reactions take place at the three-phase boundary. This model could be more accurate in terms of physical significance. The presence of liquid water provides the critical path for proton conduction. Submodel 3 describes that, catalyst surface is covered by a thin liquid film. Therefore, the solubility of H_2 and O_2 inside the film should be taken into account. Based on the three proposed models, the significance on the Nernst voltage can be assessed by calculation.

(2) Gas solubility

Submodel 3 mentioned above actually describes a state when electrode is flooded with water. Reactants need to diffuse through the thin film to reach catalyst surface. According to Nernst equation, the voltage is calculated based on the following equation:

$$E = E^\circ + \frac{RT}{nF}\ln\left(\frac{X_{H_2} X_{O_2}^{0.5}}{X_{H_2O}}\right)$$

It is clear that, the solubility actually determines the molar fraction of H_2 and O_2 in the thin film. Gas solubility values can be calculated using existing data, which is highly dependent on temperature and water properties. One can expect that, higher molar fraction of water leads to losses off Nernst voltage (Figure 4 – 13).

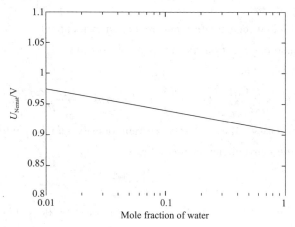

Figure 4 – 13 Dependency of Nernst voltage on water content

(3) Losses at low current

Under low current conditions, the electrochemical reaction is not carrying out at super high rates, thus ohmic and concentration polarization are not important. The losses are mainly owing to activation effects on the surface of electrode (catalyst).

(a) Reaction of dissolved gas molecules on the surface of catalysts

(b) Electron transfer between catalyst and activated molecules

(c) Transfer of ions through the electric double layer

This process has been well illustrated in previous sections (Figure 4 – 14): At anode side, H_2 molecules diffuse through the water layer to the surface of catalysts and are split into protons and electrons. At cathode side, O_2 molecules also travel through the thin water film and combine with protons in electrolyte and electrons from external circuit to finish the whole catalytic circle.

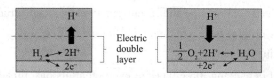

Figure 4 – 14 Mass transfer through electrolyte and electrodes

Butler-Volmer equation is often used to describe the complex dependence of exchange current density on the operating conditions such as concentration/fugacity of reactants and products. The activation overpential affect the output current density in the following manner:

$$j = j_o^* \times \left[\left(\frac{C_{reactant}}{C_{reactant,o}}\right)\exp\left(\frac{\alpha_{anode}nF}{RT}\eta_{act}\right) - \left(\frac{C_{product}}{C_{product,o}}\right)\exp\left(\frac{\alpha_{cathode}nF}{RT}\eta_{act}\right)\right]$$

Where j is the exchange current density, C represents the concentration/fugacity of reactants and products, α is the transfer coefficient for anode and cathode and n is the number of electrons involved/exchanged in the rate-limiting step.

Since the overall reaction is a redox reaction in anode and cathode, one can use a simplified model to approximate the complicated equation by assuming the following points:

(a) The surface concentration of reactants and products can be substituted by concentration of H_2 or O_2 at a single surface.

(b) Transfer coefficient for oxidation and reduction is identical.

(c) One can assume one electron transfer is involved in rate-limiting step, although the exact number might not be certain yet.

$$j = j_o^* \times \left(\frac{C_{bulk}}{C_{surface}}\right)\left[\exp\left(\frac{\alpha F}{RT}\eta_{act}\right) - \exp\left(-\frac{(1-\alpha)F}{RT}\eta_{act}\right)\right]$$

Actually, in many cases, the exchange current density can be simply represented by the following equation for the convenience of prediction:

$$j = j_o^* \times \left(\frac{C_{bulk}}{C_{surface}}\right)$$

At low overpotential condition, mathematical solution can be applied to approximate the actual exchange current density:

$$j = j_o^* \times \left(\frac{F}{RT}\eta_{act}\right)$$

Under this condition, we can introduce a new term namely charge transfer resistance R_{CT} in this equation. According to the definition, the above equation can be rearranged as:

$$R_{CT} = \frac{1}{j_o^*} \times \frac{RT}{F}$$

It is clear that, the charge transfer resistance is only depending on reference exchange current density.

4.4.5 Important variables

There are many parameters, which affect the performances of fuel cells. For example, temperature, pressure, feedstock composition, current density, cell structures, and life span of materials are regarded as critical factors influencing the cell efficiency.

(1) Current density

The effect of current density on three types of overpotentials has already been illustrated in the previous section. As expected, poor cathode kinetics causes activation overpotential. However, as current density is further enhanced, gas diffusion is a major issue. Leading to sharp decrease in cell voltage. But overall, power density is still increasing with higher current density (Figure 4 – 15). One can expect that, there exist an important trade-off in selecting suitable operating points. It is logical to design and operate cells under maximum output power.

It is interesting to observe that, the unique characteristics of fuel cells provides customers with a benefit. The efficiency is actually increasing with higher load. Even though the efficiency of other components in the fuel cell may operate at a relatively lower efficiency as the load reduced, the overall combinatory efficiency of fuel cell systems can be predictable at rather flat as load is reduced. It is in contrast with the conventional heat engines, which could operate at very low efficiency under part-load conditions.

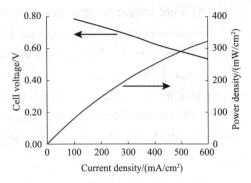

Figure 4-15 Voltage variation on current density

(2) Temperature and pressure effect

According to Gibbs free energy, the dependency of ideal potential (E) on temperature and pressure is expressed as follows:

$$\left(\frac{\partial E}{\partial T}\right)_P = \frac{\Delta S}{nF}$$

$$\left(\frac{\partial E}{\partial P}\right)_T = \frac{-\Delta V}{nF}$$

As the entropy change for H_2/O_2 is negative, increasing reaction temperature results in lowered output potential by approximately 0.84 mV/℃, assuming the product is water. As for this reaction the volume change is negative, therefore, reversible potential is enhanced with an increase in pressure.

To be more specific, temperature variation will lead to the following four changes in the system:

(a) Reaction rates. Similar to thermal catalysis, electrochemical catalytic reactions also follow Arrhenius law. Therefore, the losses are minimized exponentially with increasing temperature.

(b) Ohmic losses. Resistance of metals is enhanced at higher temperature. But for electrode and electrolyte, higher operation temperatures mean enhanced electronic and ionic conductivity. However, for aqueous electrolytes, high temperature may cause dehydration thus resulting in poor conductivity. In general, high-temperature fuel cells display higher efficiency as resistance is reduced. But for low-temperature fuel cells the impact is limited.

(c) Mass transfer. Compared with electrode reaction, the enhancement of mass transfer is limited, as the activation energy for diffusion coefficient is much lower than kinetics, in general.

(d) High operating pressure actually exhibits several benefits on fuel cell performances. This is because that partial pressure of reactants, solubility in electrolyte and mass transfer rates are all enhanced under pressurized conditions. However, higher pressure also means increase in equipment cost, as thicker piping and reactor are needed with additional expenses in manufacture and maintenance. For high-temperature fuel cells such as MCFC and SOFC, pressure difference should be handle with great care because gas leakage through electrolyte and seal could lead to hazard formation. Pressure also impact reaction pathways for natural gas conversion. Higher operating pressure might accelerate coke formation.

(3) Fuel utilization and gas composition

According to Nernst equation, the fuel and oxidants with higher partial pressures lead to higher cell voltage. We define such efficiency as utilization, referring the fraction of total fuels or oxidants introduced into the cell, which react electrochemically. In low-temperature fuel cells, determining the fuel utilization efficiency as the ratio of amount of H_2 being converted to that being introduced. Similar definition could also be applied to oxidant (O_2). For example, in MCFC devices, both O_2 and CO_2 are used for electrochemical reactions.

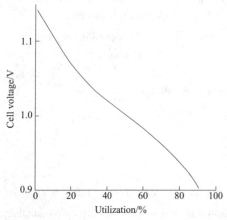

Figure 4 – 16 Cell voltage as a function of CO utilization

When CO is used as fuel, the fuel utilization efficiency could exceed theoretical value of H_2 utilization efficiency. This is because CO reacts with H_2O to produce more H_2 and water during shift reactions (Figure 4 – 16). This reaction reaches in equilibrium rapidly under elevated temperatures. For example, for inlet fuel composition of H_2 (30%) and CO (10%), fuel efficiency can be very high, and H_2 utilization efficiency could be higher than 100%. To better describe the fuel utilization, the fuel efficiency should be defined as the ratio of H_2 consumed to the sum of fresh H_2 and CO introduced at the inlet of the cell unit.

As already mentioned in previous section, high reactant utilization efficiency results in decreased output voltage. A representative graph well illustrate the variation of cell voltage as a function of reactant utilization efficiency. Increased utilization from 20% to 80% could cause a decrease in output voltage of approximately 0.16 V.

4.4.6 Modeling

Mathematical modeling is critical for fuel cell development (Figure 4 – 17). Because modeling work helps reveal the behaviors of fuel cells, allow optimization of operating conditions, minimize the materials and thus enable optimal performances and economics of fuel cell devices. Modeling is particularly important for fuel cells due to the uniqueness and complexity compared with conventional internal combustion engines. A few types of models have been applied to elucidate the behaviors.

Internal physics and chemistry of fuel cells. Experimental studies and characterization is often very difficult in actual systems. So models can help understand the critical processes in fuel cells. Mathematical models can also help focus on experimental development efforts. Models can be used to guide researchers to improve the experimental performances and improved interpolation and exploration of experimental data. Rigorous modeling can also help researchers to find the right position of scientific hypothesis and provides reliable framework for testing those assumptions.

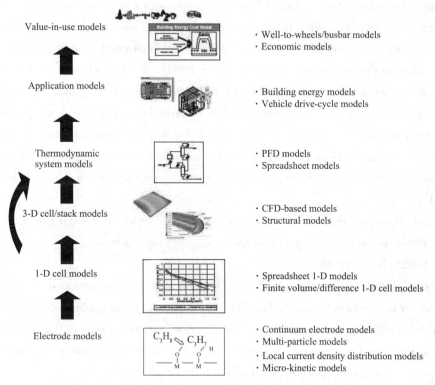

Figure 4 – 17 Mathematical modeling for fuel cell technology

Modeling can also help support system design and performance optimization. Fuel cell systems have so many units and components. Thus, designers should find a suitable system model which are critical for effective system design. R&D for fuel cells need to fully consider dynamic models. For these systems, modeling can also help evaluate the technical and economic performances of fuel cell in actual applications. Those models are used to determine whether the performance of fuel cells are in the optimized condition, which will match the requirements of a given requirement from the customers. Based on the discussion above, it is clear that each of these applications for cell models should have a specific requirement in terms of the level of details and rigorous in modeling processes. Thus, it should be predictive for the capability of fuel cell performances.

In many high level applications, the predictive requirements is very important. But in some cases, the operational characteristics of fuel cells, is not even an important factor. Simplified models are satisfactory and should be appropriate for general customers. In most daily use, it is possible to encapsulate the mass/energy balances and performance equations for fuel cell within a special application. Such simplified models are satisfactory for quick trade-off considerations.

In addition, developing models can also help to improve understanding of complex physical and chemical phenomena and to optimize unit geometries to improve the overall efficiency of fuel cells.

As already discussed above, given the consideration of a wide range of potential applications, such as in stationary units, transportation vehicles and portable devices. Despite of the wider availability of quite sophisticated fuel cell models, researchers and engineers for fuel cells must be quick users. Because obtaining reliable experimental data on the behaviors of fuel cells, particularly for different systems, can be very difficult time consuming and cost ineffective. With decades' effort for the fundamental research, unfortunately, this has not been reach to satisfactory and accurate stages where a detailed data of sufficient quality and quantity allows thorough, validation of mathematical models. Numerous data reported in literature on future performances is interesting but insufficiently accurate and not suitable for developing reliable models. In fact, a lot of data obtained in cell and stacks structures were taken at a modest utilization. Therefore, it is almost impossible to interpret this data and reliable kinetic data model, thus cell model cannot be developed.

Based on the above discussion, it is clear that the developers of the cell model must be very familiar with the assumption limitation, as well as experimental details.

(1) Value-in-use models

This type of model is a mathematical model, which allow users to predict how unique features of fuel cells create additional value or benefits for a particular given application. Therefore, such type of model is highly application-oriented. For example, a typical model of this type would be an economic model, which assists users or developers to predict the cost resulting from the installation of a device in buildings and in other residential applications.

This type of model is also well-known to assess the energy consumption, environmental impact and sometimes the cost of different transportation options used in vehicles. Actually, this type of model is commonly used to evaluate the performance of PEFC devices, which is another important application. This type of model can be used to help manufacturers to establish cost and performance relations for fuel cells.

(2) Application models

Also, application models have been developed to assess the interactions between fuel cell power system and the user environment. The most common use in vehicle applications is that, some dynamic interactions exist between the power system and the operation environment for the vehicles. Because it is too complex to analyze without help of a mathematical model. Several commercial software has been developed to predict the dynamic vehicle behaviors. The users can evaluate the vehicle specific specifications, such as electric motors, batteries, engines, and connections. Industry partners can also contribute to develop more state-of-the-art algorithm to ensure the accuracy of the application models.

(3) Thermodynamic models

This type of model is developed for system analysis just as process industry where many software such as Aspen Plus, HYSYS and have been used to simulate the process parameters using thermodynamic models. This type of model has been used routinely by fuel cell researchers and developers. This is a very powerful tool for system engineering. Selection of appropriate thermodynamic model is critically important in this case. Overall accuracy is often very good. Because fuel cell models or stacks are typically lumped parameter models, or simply sum-up in tables.

Thermodynamic models were originally intended for use in applications such as stationary power generation in order to optimize process performances and to evaluate performance parameters. Therefore, a large library of unit operation models is needed to simulate the process unit operation and different equipment. It should also has a library of various types of chemical components and data to predict the physical and chemical properties, which could be used to calculate thermodynamic features. As operation deviates from the set point conditions, for example, a reference date, a voltage signal or adjustment should be applied to tune the perturbation. Obviously separate voltage signals or adjustments can be applied for current density, temperature, pressure, and fuel utilization efficiency as well as for fuel composition and other operation conditions.

(4) Stack models

This type of model can be used to estimate the geometry effect of cell and stack to help engineers to understand the impact of stack operation conditions. Considering the variety of different possible stack geometries and inter connects as well as operation parameters such as temperature, pressure, and component concentrations. It is very difficult to perform optimization of set design without a fundamental understanding of the stack models. Therefore, it is essential to develop a model which represents the characteristic physical and chemical features of a stack. In addition, the way of interconnect and a number of different three dimensional stack models should be developed for this purpose.

To develop reliable models, the stack geometry should be discretized into finite elements, which can be assigned with the appropriate properties of materials and components. At the same time, the models should also represent the intrinsic kinetics of electrochemical reactions. The following features are often included in stack models (Figure 4 – 18).

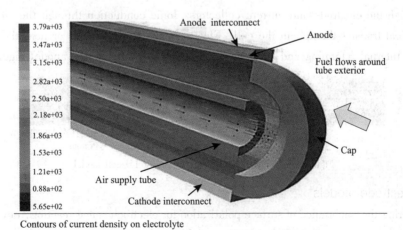

Contours of current density on electrolyte

Figure 4 – 18 Tubular stack model

Computational fluid dynamics-based fuel cell codes. Commercial see CFD codes can be modified to represent the electron electrochemical reactions, electronic and ionic conduction, as well as stack geometries. Many cases, refinements in treating the chemical reactions and flow pattern through the porous material should also be incorporated to better represent electrode processes. In addition to assessing basic performances, this model can also help understand the impact of different manifolding arrangements.

Computational fluid dynamics based fuel cell codes. The commercial CFD codes can be modified to better represent electrochemical reactions and electronic and ionic conduction. In many cases, refinements should also be conducted to better represent both reactions and the stack geometries, as well as flow patterns through porous media. In addition to assessing the basic performances for fuel cells, such as current density, temperature, and concentration profiles. These models developed should also help understand the impact of different manifolding arrangements.

Computational structure analysis. These codes are developed based on commercial available 3D structure analysis calls such as ANSYS. Typically, these models should be augmented to represent an ionic conduction, fluid flow patterns as well as chemical reactions and kinetics. However, these models do not provide more insights into the complex flow patterns. They are actually very useful to evaluate the structure robustness. For example, these models can be used to assess mechanical stresses in the stack, which is a key issue in some of the high temperature fuel cell technologies.

CFD analysis computes local fluid velocity, local pressure and temperature throughout the region of interest. By coupling CFD predicted behaviors with the electrochemistry, as well as thermodynamics, detailed predictions are possible for researchers to improve the model reliability.

(5) 1D Cell model

In 1D models, all of the critical phenomena, such as reactions and transport can be considered in 1D fashion. In general, the following elements should be incorporated:

Convective mass transfer of reactant and products to and from the surface of the electrodes. Mass transfer process of reactant and products through the porous media. Conduction of electronic current through the electrode and current collectors. Ionic conduction through the electrolyte and electrode. Heat transfer throughout the cell. Electrochemical reactions at or near the three-phase boundaries. Internal reforming and shift reactions taking place inside the anode (Figure 4-19).

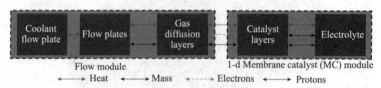

Figure 4-19 Schematic description for 1D cell model

(6) Electrode models

Considering the importance of surface polarization for electrochemical reaction, developing good electrode sub models is critical in developing reliable fuel cell models. The following four scale approaches should be employed to develop different levels of electrode models (Figure 4-20):

Continuum electrode approach. In this model, electrode is considered as a homogeneous medium for diffusion, electrochemical chemical reactions and ionic conduction. Actually, this approach neglects the specific process occurring during the diffusion of reactants and products

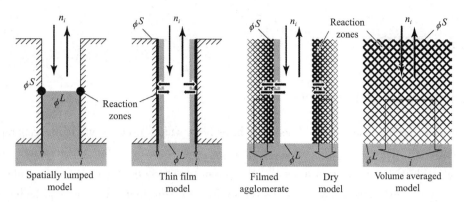

Figure 4-20 A scheme for electrode model

through the electrode. This model cannot distinguish the difference between the rate limiting steps involved in a surface reactions and diffusion limited processes.

Multi particle approach. This model recognizes that electrodes are actually in a different phase from the reactant and products. Therefore, issues such as connectivity and interfacial mass transfer related factors are addressed within this model. This model is more like a hybridization of the different continuum models.

Micro-kinetics approach. In this approach, the individual reaction steps or the elementary steps at or near the surface are considered. Generally, this approach are usually imbedded into the multi particle or other models which can better represent the actual flow pattern and reactions to predict the behaviors of fuel cells. This is only approach that can give more insights into the rate determining electrical chemical reactions, which is taking place in the cell.

(7) Obtaining data from single cell experiments

Single cell experiments are often carried out to obtain basic data for stack performances. Particular attention should be paid to check and validate the reliability of experimental data and those obtained from literature. Important parameters such as current density, flow rate, temperature, flow field configuration should be carefully checked prior data fitting. In general, single cell model development should follow standardized procedures to obtain parameters from a polarized curve.

(a) Obtain ohmic resistance from EIS experiments or literature of identical materials.

(b) Subtract the resistance loss from the polarization curve, thus Tafel plot in low current region can be used to assess activation polarization.

(c) Compare the output of the developed model with experimental polarization to obtain transportation significance.

As already discussed in previous sections, as the starting point, the following equation is often used for computing the activation loss through Tafel curve.

$$V_{cell} = V_{cathode} - V_{anode} - iR - \eta_{con}$$

By fitting the model for Nernst voltage ($V_{cathode} - V_{anode}$), exchange current density can be calculated and compared with literature data.

4.5 PEFC

4.5.1 Cell components

Polymer electrolyte membrane. Fuel cells have been widely used to efficiently generate high power densities, thereby making this technology very attractive for mobile and portable applications. PEFC has been used as one of the prime power energy technologies for automobiles and have captured extensive attention across the world. Different from other fuel cell technologies, PEFCs use solid phase polymer membrane as cell separator and electrolyte. Since this type of cell is operated under mild temperature compared with other cells, typical issues such as sealing, assembly and handling is not as significant as high temperature fuel cells. More importantly, PEFCs do not need to deal with waste acids at low temperature (<100℃) as solid acid membrane is used to avoid environmental hazards. Thus, PEFCs are regarded as most promising candidates for future fuel cells. As extension of H_2 fuel cells, methanol, ethanol, propanol and formic acid based fuel cells have also been developed potentially for commercial uses.

In PEFC structures, ion exchange membrane, conductive porous layer, electrode/electrocatalysts between the layer and membrane, cell interconnects and flow plates for planar bipolar type of cells.

As depicted in Figure 4 – 21, in H_2/O_2 PEFCs, H_2 molecules diffuse through the backing layers towards the anode, where electrons and protons are generated. Protons then travel through the membrane to reach cathode for O_2 reduction reaction. Similarly, O_2 also needs to diffuse across the backing layer for sufficient contact with protons and electrons. Gas vents are connected with consequent units to ensure continuous flow in the stack.

The main components for single cell structure are presented in Figure 4 – 21. Carbon block (graphene), Teflon mask are needed to seal the gas and ensure gas flow. Catalysts for H_2 oxidation and O_2 reduction reactions can be manufactured in a layered form and attached to two sides of membrane. In other words, proton membrane can be sandwiched with cathode and anode catalytic materials. Each of the main components in PEFC will be briefly discussed in this section, detailed of which will be illustrated in later sections.

(1) Membrane

Original acidic fuel cells were designed using H_2SO_4 as electrolyte in aqueous medium. Organic based cation exchange materials were developed for fuel cells in 1960s, to avoid generation of waste acids and severe corrosion problems. This initial effort had led to considerable efforts in developing perflurosulfonic acid polymers for today's systems. The main function of ion exchange membrane is to provide ionic path in solid phase materials, while separating fuel and oxidant gases at the same time. Teflon is an electric insulator while ion conduction occurs *via* ionic groups inside the polymer structure.

PEFC devices have been long developed for automobile and stationary applications. PEFCs

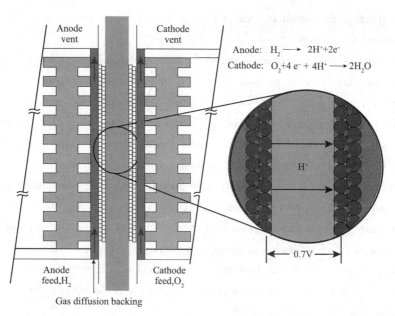

Figure 4−21 Configuration of PEFC

are constructed through membrane electrode assemblies (MEAs). The single cell unit in laboratory is similar to the units in actual applications. Typical commercial devices can be operated for over 20000 h continuously with 4 ~ 6 μV/h degradation rate. MEAs have been implemented for public transit buses and passenger automobiles. For stationary applications, the power output is typically around 2 ~ 10 kW. But stack life is the limiting factor in MEAs. Commercial PEFCs cannot sustain over 1000 h continuous operation due to fouling of stack materials. This is a major cost contributor for both transportation and stationary applications.

The most well-known brand for fluorinated Teflon family materials is Nafion. It has a typical equivalent weight range of 800 ~ 1100 milliequivalents per dry weight of polymer. Such membranes exhibit remarkable chemical and thermal stability against chemical corrosion in strong alkaline and oxidative environment. Nafion materials consist of fluoropolymer backbone similar to Teflon, with sulfonic acid groups chemically bonded inside the channels of polymers. In most fuel cell and electrochemical systems, Nafion has a lifetime of 50000 h for demonstration.

Melt extruded manufacture approaches are traditionally used for Nafion products. Solution based film processes have also been emerging as alternative methods to reduce the overall cost of Nafions. Mechanical properties have also been improved by employing internal support layers.

(2) Porous backing layer

The polymer membrane is actually sandwiched between two porous backing layers. Such layers can provide mechanical support to prevent membrane from breaking. The functions of backing film include (a) gas diffusion medium, (b) electrical pathway for electrons, (c) gas channels as well as collectors and (d) water removal from electrode. Typically, the porous backing materials are made of carbon-based substrates, or cloth form with pressed carbon fiber configuration. The layer is incorporated with hydrophobic functional groups such as PTFE to prevent water flooding the membrane and electrode materials, to ensure gases diffuse freely to contact the surface of catalysts.

Water removal is one of the most important functions for backing layers. It is generally believed that, two mechanisms possibly determine this process. transport of water molecules through the porous plate into coolant and evaporation into reactant gas streams. Graphite-made layers can direct formed water with coolant and eventually to the reservoir. Smooth removal of water is important, as additional liquid layer would create additional mass transfer resistance for reactant gases to diffuse across the surface of electrocatalysts. But sufficient wetting of such layer is also important, as it would supply enough water to humidify incoming gases thus preventing drying of membrane, which is a common failure mode.

Other advantages of removing water include the reduction of downstream exhaust condenser for oxidant, and improved fuel utilization efficiency, at elevated temperature which also contributes to simplified control system. In fact, the backing layer is critical to control the climate inside the stack. Therefore, achieving optimal water vapor pressure or humidity is a challenge in fuel cell operation.

(3) Electrochemical catalyst layer

The backing layer is in close contact with membrane, while the electrodes/catalysts need to be in high intimacy with backing layers to ensure smooth proton mobility. It is a critical factor determining the observed performances of electrochemical catalysts. The so-called "binder" is important and performs multifunctions, such as fixing catalyst particles within the layered structures, and acting as electrode architecture bearing the performances.

Two main mechanisms are accepted to govern the catalyst or binder design. PTFE materials within the electrode can prevent full wetting to ensure efficiency of surface reaction. Another mechanism is that hydrophilic electrode could enhance membrane-catalyst contact, thus the overall loading of Pt content in electrodes can be minimized to large extent. Actually, in many MEA devices, catalysts are embedded in the structure of polymer membrane, in order to gain high solubility of protons and O_2 molecules on the surface Pt sites.

Pt metal is used for both anode and cathode catalysts. For other types of fuels with CO as impurity, bimetallic PtRu catalysts are usually desired to prevent poisoning of Pt metal. The size is important for utilization of noble Pt metal. Many attempts have been paid to designing Pt-based alloys with other metals such as Au, Cu, Co, Ni, to reduce the cost of noble metals.

The content of Pt metal is as high as 40% on carbon support. Alternatively, active Pt metal can also be deposited on the surface of porous backing materials, or directly bonded to membrane. High Pt loading (~ 4 mg Pt/cm^2) could contribute to a 600 mA/cm^2 density. Operating temperature is often below 90℃. MEA performances can be sustained for over 10000 h.

Low temperature operation can be both advantageous and disadvantageous. Pure H_2 is highly needed as ppm level of CO contaminated fuel would lead to severe poisoning of Pt metal. In recent years, alternative membrane materials such as polybenzimidizole (PBI) are developed as good candidates for PEFC operated at slight higher temperature above 120℃, where dehydration of membrane remains a grand challenge. But PBI only requires small amount of water to accelerate proton transfer. Thus water management is dramatically simplified. The only issues still plaguing with such alternative membrane is that, corrosive phosphoric acid needs to be incorporated inside the membrane to provide sufficient ionic conductivity.

Higher operating temperature also means generation of heat with higher value, which is important for stationary applications.

4.5.2 Electrode materials

(1) Basic structures of electrocatalyst

MEA structure is the heart of PEFC devices (Figure 4 – 22). As mentioned in the prior section, catalyst materials are actually doped on the backing layers which sandwich membrane in the middle. Several catalytic materials can be deposited as needed. The most important feature is that, catalytic active sites should be close contact with layers and membrane structures. Selection of suitable anode and cathode catalysts is very important. For anode catalysts, materials which cannot sustain highly oxidative environment, or use for O_2 evolution reaction, are not selected as anode catalysts.

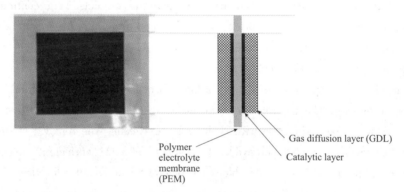

Figure 4 – 22 Polymer membrane

The theoretical output potential for fuel cells is 1.23 V. The stability of both anode and cathode materials is equally important. For cathode catalyst, since ORR (O_2 reduction reaction) is much slower, kinetically, compared with H_2 oxidation reaction, improving the long-term performances under acidic/alkaline condition remains a grand challenge.

(2) Preparation of electrocatalyst

Polyol method. Reduction of Pt precursors using polyols has been widely studied for acidic fuel cells. $C_2 \sim C_4$ polyols with varied boiling points from (197 ~ 290℃) are used to promote the nucleation and growth of nanoclusters. Since the oxidation potential of each polyol is different, reduction temperature is set considering the nature of solvent. Ethylene glycol is the most commonly used solvent for synthesizing Pt electrocatalysts (Figure 4 – 23). Because ethylene glycol has

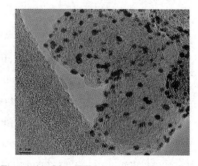

Figure 4 – 23 TEM image for Pt catalyst prepared by polyol method

a high enough boiling point and high dielectric constant. It is also important to mention that, pH is often tuned at 11 ~ 13 to stabilize the nanoparticles.

Earlier studies have already confirmed the mechanism underlying the formation of evenly distributed Pt particles in ethylene glycol as solvent. Metal hydroxides are believed to form first in alkaline medium. In this mechanism metal acetaldehyde was considered as the most important intermediate for eventual formation of metal clusters. Oxidation of ethylene glycol to glycolic acid also plays an important role, as the latter can stabilize the cluster and manipulate the growth rate of nucleus. Therefore, size and stability of Pt nanoparticles can be largely determined by the presence of carboxylic acid and hydroxide ions in aqueous medium.

Microemulsion method. Microemulsion actually represent macroscopically homogeneous fluid made by mixing aqueous and organic phase in the presence of surfactants. Even the fluid looks homogeneous, actually at nanoscale or even microscale, it is indeed consisting of solution with mono-sized droplets. Therefore, the size of nanoparticles is strongly determined by the size of these droplets.

Two types of system are popular for microemulsion synthesis. For oil-in-water (o/w) system, the internal structure of emulsion is composed of small oil droplets in a continuous aqueous medium. The surfactant molecules are oriented with hydrophilic portion towards aqueous medium. Under such condition, the size of oil droplets is difficult to control because metal precursors are actually dispersed in aqueous medium.

Water-in-oil (w/o) type of emulsion has a high oil concentration and internal structure with water droplet as the core. Clearly, the size of metal nanoparticles is determined by the size of the droplets. But there is one main issue with such biphasic system. The removal of surfactant from the metal surface cannot be well achieved. As a matter of fact, surfactant molecules interact strongly with metal particles thus possibly blocking some of the active sites for electrocatalysis. For w/o type of emulsion, two micelles are prepared, one with reducing agents such as $NaBH_4$ or hydrazine in aqueous solution, and one with metal precursors. Mixing of the two micelles leads to collisions among droplets and formation of metallic seeds. Nucleation step also occurs inside the droplet. Surfactant molecules control the growing processes of metal particles through steric effect. Selective and strong adsorption on preferable metal facets is the intrinsic driving force to eventually form target size and shape.

To immobilize the as-formed nanoparticles in homogeneous medium, solid supports are then introduced and mixed with the medium to facilitate metal dispersion on support. The catalysts need to be washed and cleaned completely to eliminate surfactants on catalyst surface.

Impregnation-reduction method. This is one of the most popular method in preparing Pt based solid catalysts, as this method is relatively simple and cost effective in terms of materials and energy used during impregnation and activation. The support can be immersed inside an aqueous solution containing metal precursors. Various metal precursors such as chloride, sulfite, nitrate and carbonyl complexes can be effectively reduced into metallic component. The second step is reduction, where metal precursors are reduced in the presence of H_2 or reducing agents similar to those used in microemulsion method. The main flaw with this method is the difficulty in controlling the homogeneous distribution of size and shape without use of surface directing agents (surfactants).

Other methods. Various other approaches have also been used for preparation of Pt catalysts. For example, KBr is an important size-controlling agent (Figure 4 – 24). Because the large size of bromide anions ensure the steric hindrance for metal cations. It prevents the agglomeration of metal particles in aqueous phase. Concentration of metal precursors, Br^-/precursor ratio and temperature are critical factors.

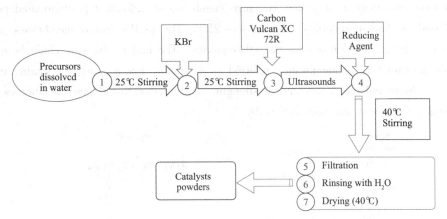

Figure 4 – 24 Description of KBr promoted catalyst preparation method

In a typical process, Br^-/precursor ratio is set at 1.5 with 10^{-3} M of metal precursor at room temperature. Low metal precursor concentration means that large flasks are needed for synthesis. In another approach, precipitation of metal precursors in the form of hydroxide or carbonate can be carried out prior to instant reduction by reducing agents.

4.5.3 Membrane materials

(1) Membrane structure

The advantages for using solid phase polymer as electrolyte have been illustrated in previous sections. Membrane materials need to meet the following requirements:
- high ionic conductivity.
- impermeable for gases and electrons.
- high tolerance under corrosive environment.
- high mechanical durability.

To meet the above standards, sophisticated design of polymer structures needs to be carefully conducted. The functional groups, or acidic groups are considered as the most important parameters affecting physicochemical properties of the ion exchange membrane. The functional groups consist of fixed group and counter ion. When the membrane is hydrated with water, the functional groups will dissociate to release counter ions for ionic conduction. Based on the charge of counter ions, ion-exchange membranes are categorized into two types:

(a) Cationic exchange membrane. Negatively charged fixed groups with cations as free ions.

(b) Anionic exchange membrane. Positively charged fixed groups with anions as free ions.

In PEFC devices, cationic exchange membranes are widely used. To be more specific, proton exchange membrane fuel cells, namely PEMFC are commercialized products. They are based on sulfonated aromatic hydrocarbon or perfluorinated polymers. DuPont and Solvay have developed perfluorinated ion exchange membranes. Hybridized perfluorinated membranes with SiO_2 or ZrP have also been implemented in applications. Sulfonated hydrocarbons or poly(trifluorostyrene) are also widely used in many PEFC units.

The most commonly used proton exchange membrane is sulfonated perfluorinated polymers, often referred as Nafion materials (Figure 4 – 25). The perfluofinated membranes of Nafion materials have long side chains with two ether groups. The end of the side chain is sulfonated group. For another type of membrane, referred as Aquivion, has a short side chain, making it possible to achieve enhanced mechanical strength. Hydrocarbon backboned membranes are still under development for commercial fuel cells.

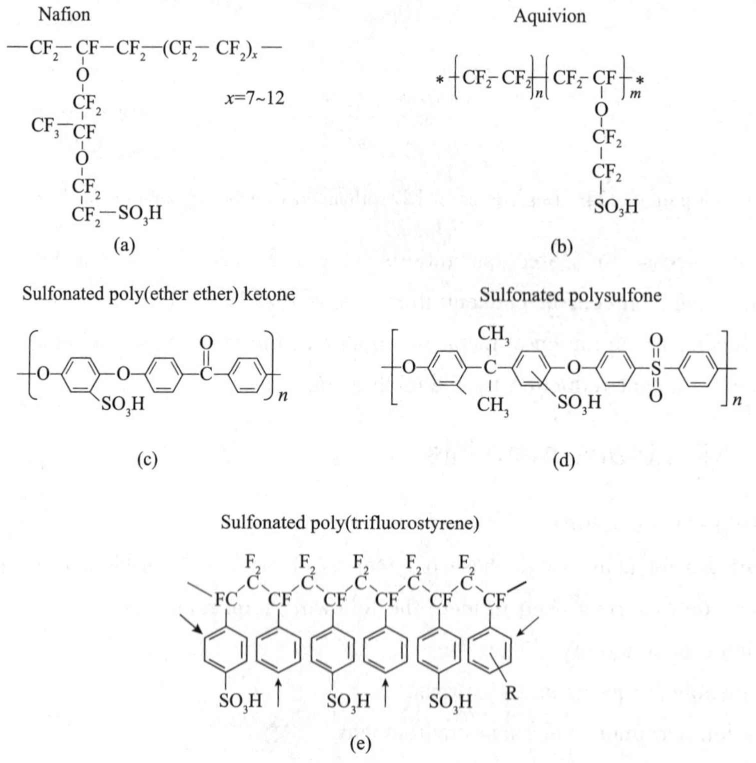

Figure 4 – 25 Common proton exchange membrane units

For future R&D of membrane products, the following new characteristics are increasingly being considered for better performances of PEFC products.
- Hybridized SiO_2 or ZrP to improve mechanical and morphological properties.
- Chemical modification of morphologies using plasma or radiation techniques.
- Synthesis of novel polymer electrolytes from waste or low-value materials.
- Immobilization with other ionic conductive chemicals such as ionic liquids.

(2) Properties and equations

Important definition. Both physicochemical and transport properties are very important for membranes. Numerous physicochemical parameters need to be carefully considered, including thickness, density, tensile strength, swelling ratio, equivalent weight, ionic conductivity, water uptake, hydration number, water sorption isotherm, diffusion permeability, electroosmotic permeability. Typical film thickness is 7~25 μm. Swelling factor describes the relative change in membrane size after swelling in water. For Nafion membrane, the swelling factor is approximately 15%. Water uptake is defined as the water content in the membrane after absorbed with water. Equivalent weight of polymer fragment containing one functional group is measured in g/mol. Hydration number is defined as the amount of water molecule per one functional group in membrane, which is in mol H_2O/mol SO_3^-.

Water sorption isotherm characterizes the dependence of membrane hydration number on water vapor humidity. Fundamental understanding on this behavior is essential for strategizing water management (Figure 4-26). It is often used to establish mathematical modeling.

Obtaining the isotherm has been extensively studied. Particular attention should be paid to the difference between the equilibrium of membrane with saturated water vapor and liquid water. Prediction of the equilibrium for water vapor and absorbed liquid water molecules in the presence of strong water-functional group interaction is still very difficult, thus some researchers have measured and proposed some empirical models.

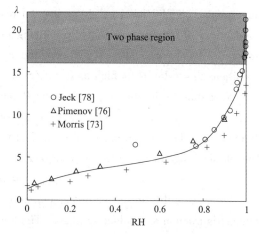

Figure 4-26 Hydration number on water vapor humidity

According to experimental results, the maximum water concentration in membrane increases with higher temperature. Water content in the system causes dramatic rise of the membrane water content.

At higher relative humidity, water vapor might be condensed on the surface of membrane. However, only a few water droplets will not be enough to increase the overall water concentration.

Ion channels. Structural characterization already confirmed that, hydrated functional groups in membrane actually formulate clusters consisting of three or four hydrated groups as inversed micelles. The side chains actually dangle inside the channels and free ions diffuse through the tunnel. The cluster-network model has been studied to predict the proton transport properties. It is believed that, the conductivity of membrane is characterized by the motion of ions. Another mechanism proposed that, H bond formed in aqueous water from O atoms of water and free protons governs the transport rate of ions through the membrane.

The hydrophobicity of channels determines the proton mobility. Experimental studies have demonstrated that, hydrated protons travel much faster in perfluorinated structures than in hydrocarbon membranes. Computational results also showed that, the strong OH···O bond between each water molecule is the main driving force for formulating hydrated protons. The results confirmed that, H bond distance is 1.85 A in average. The perfluorosulfonated membrane has a higher water binding energy than aromatic membranes. The key reason is that perflurosulfonated membrane actually makes dissociation of sulfonyl functional groups much easier than hydrocarbon aromatic membrane.

Water transport and conductivity. The volumetric fraction of conducting phase determines the conductivity. The fraction of conducting phase is dependent on the concentration of sulfonyl group in the membrane. The relative humidity of membrane environment affects the total water amounts in the channels. As water concentration grows up, both volumetric fraction of conduction phase and water diffusion rate are enhanced. It also causes increase in electro-osmotic coefficient, thus proton conductivity rises. Experimental results have shown that, both diffusivity of water and conductivity of ions grow directly with rising hydration number. It is important to mention that, under dry condition, the specific conductivity of membrane is 10^{-6} S/m, while in hydrated conditions this value is as high as 0.1 S/m.

According to the theory proposed by Broadbent and Hammersley, the fraction of conducting phase (f_{cr}) needs to reach the critical value, after which the conductivity is enhanced dramatically. The transformation from dielectric to conductor is referred as percolation transition. The physical significance can be described as follow:

$$\sigma_{membrane} = \sigma \times (f - f_{cr})^\tau$$

Where σ is the pre-factor of percolation equation and τ is the percolation exponent. f is the volumetric fraction of conducting phase. The following graph (Figure 4 – 27) can also describe the change from dielectric to conductor for the membrane. Once the functional groups are hydrated, hydrophilic groups are clusters and ionic path/channel is formulated, the circuit is closed and energy output is achieved. Based on theoretic calculation, $f_{cr} = 0.15 \pm 0.03$, $\tau = 1.6 \pm 0.4$.

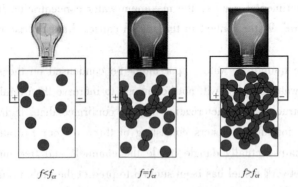

Figure 4 – 27 Change from dielectric to conductor for the membrane

Bipolar plates. MEAs are the most important components in PEFC stacks. They are essential for many functionalities in fuel cells, (a) providing sufficient mechanical strength to hold the structure of stacks, (b) and electric conductivity, (c) heat management and prevent water flooding and ensure instantaneous removal. Thin plates can assist reducing the overall size and weight of the MEA stack. In addition, good resistance against corrosion is important. According to the weight contribution from all components in the cell, bipolar plates actually account up to 81%~88% of total weight, while gasket, membranes, electrodes, and current collectors are only 15% in total weight.

Biopolar plates can be classified into two types based on the constructing materials. The first type is graphite or graphite-polymer composite material. Graphite materials have the advantages such as low contact resistance, and good durability in acidic environment. However, graphite materials are brittle. Therefore, polymers are often incorporated with graphite in mass manufacture. For example, polypropylene (PP), polyphenylene sulfide (PPS), polyvinylidene fluoride (PVDF), and phenolic resins can be utilized as binder materials. The mixing and distribution of conductive carbon within polymer matrix is of great importance for reducing electric resistance and minimizing the gas leakage, particularly under high vibration working environment.

The second type is metal based plates. This type of materials shows high mechanical strength and good electrical and thermal conductivity. Moreover, easier fabrication of flow-fields makes metal based materials very attractive for PEFC applications. However, poor resistance against corrosion is a major issue.

MEA stacks are often affected by leached metals from electrocatalysts and membrane fouling become severe under acidic environment. Surface passivation of metal materials to metal oxide also unfavorably increases the contact resistance. Metal based plates are often made of aluminum, stainless steel, titanium and nickel. Carbon-metal composites have been developed to prevent surface passivation and electronic insulation. Electronic deposition and vapor deposition methods are employed to coat nitrides and carbides.

For flow-field design, flow distribution and pressure drop undoubtedly influence the cell performances. Flow direction, channel shape/length and configuration determine pressure, temperature and humidity.

4.5.4 PEFC Systems

(1) Direct H_2 PEFC

This type of cell requires extensive thermal and water management to ensure PEFC stack operated at desired conditions. Key components in H_2 PEFC include heat exchangers, humidifiers, condensers. For automotive applications, the design standards demand that the engines should work at approximately 60℃. Thus, the temperature difference between working medium and coolant is very small. As a result, water management need special attention as significant amounts of water might be loss in the vent if operation is in an arid region.

Another critical part in H_2 PEFC is the storage tank. H_2 storage materials are considered extensively ranging from pressurized tank to metal hydrides and chemical compounds. Each of the storage options need detailed design and has its own advantages. But each option may need to compromise the energy density, weight, cost and energy efficiency against others. Overall, size and weight are important factors for the system. At present, additional reduction of size and weight of PEFC need to be carried out to show more efficiency competitiveness against traditional internal combustion engines.

(2) Reformer-based PEFC

The main advantage of reformer-based PEFC is the reduction of expensive H_2 storage parts. But the system needs to be design to process hydrocarbons. The following additional units will be incorporated.

(a) Fuel preheating and vaporization. Different from H_2 PEFC, hydrocarbons need to be preheated and vaporized to meet the requirement of reformer, in order to generate enough H_2.

(b) Reformer. Two oxidants are usually used for reforming reactions, air and steam, namely partial oxidation and steam reforming, respectively. Typically, partial oxidation reformers are smaller, cheaper and respond faster. Steam reformers ensure a relatively higher system efficiency.

(c) Water gas shift reactor. This reactor converts hydrocarbons/alcohols with water into CO_2 and H_2. It is critical as the purity of H_2 is important for PEFC stack performances.

(d) Purification. Trace amounts of contaminants are detrimental to fuel cells. CO and sulfur compounds are problematic and must be reduced to 1~10 ppm.

The additional unit operations weight and volume, thus the overall efficiency of PEFC is typically ranging from 75%~90%. The following issues further affect the structure of the whole units.

(e) Different from H_2 PEFC, H_2 generated from reformer is often diluted with CO_2 and inert (N_2). Therefore, the concentration of H_2 in the cell is lower than 30%. It decreases the ideal potential of the cell and increases concentration losses.

(f) More Pt catalysts ($0.4 \sim 1$ mg/cm^2) should be used to prevent the poisoning caused by CO and sulfur compounds. But, the power density is still 30%~40% lower than H_2 PEFC.

(3) Methanol PEFC

Structurally optimized PEFC can also process methanol as the fuel, known as direct methanol fuel cells (DMFC). Using methanol could lead to several advantages, such as mild operating temperature, high energy density as fuel is in liquid phase. Performances of DMFC can achieve $180 \sim 250$ mA/cm^2. The cell voltage is not high resulting from the cross over of methanol from anode to cathode side. Overpotentials at anode is also significant, because of the strong adsorption of intermediates of decomposition products from methanol. Very high overpotential at cathode is caused by poisoning of cathode catalysts.

Unfortunately, the catalyst loading for Pt is almost 10-fold higher than that in H_2 PEFC. The cross-over of methanol is severe if concentrated methanol is fed to the cell. The cross-over percentage is as high as 30%~50%. But decreased concentration of methanol could also lower the current density due to reduced activity of catalysts. Therefore, development of more advanced electrolyte materials, which prevent methanol cross-over are urgently demanded in the short future.

Another issue is the water transport, as methanol solution is one anode side while air is on the other side. Therefore, the water management is also very difficult in DMFC systems. Complex water recovery or management solutions are highly demanded, which may complicate the simple design of DMFC systems. Therefore, DMFC has very limited applications in both automobiles and stationary units, unless the cross-over issue is solved carefully. Special MEAs are developed to reduce water cross-over. The power output is in the range of 1 W ~ 1 kW. Electrocatalysts have also been under extensive studies. PtRu bimetallic catalysts have demonstrated to be remarkable at 200 mA/cm^2 at 0.3 and 80℃.

4.5.5 Applications

(1) Transportation

Up to date, PEFC is the only type of fuel cell considered as prime motive power for on-road vehicles. PEFCs fueled by H_2, methanol, hydrocarbons have been integrated with light duty car and trucks by almost twenty automobile companies. The following parameters are limiting ones which could possibly prevent large implementation of PEFC technologies.

The overall size and weight of PEFCs should be further reduced. Durability of membrane and electrocatalyst should also be improved. The structural design should be optimized in order to sustain the extreme working environments expected for vehicles. H_2 stations should be in rapid infrastructure to meet the market demands. Safety protocols must be developed by the industry.

(2) Stationary units

PEFCs are developed for stationary units to operate on natural gas or propane. Significant efforts and progresses have been achieved to implant such units for stand-alone applications. The system efficiency still requires further improvement (~30%). System life has been extended to approximately 8000 h.

4.6 AFC

Alkaline fuel cells (AFCs) have been largely implemented for US space programs. Desirable attributes of AFCs include active cathode ORR kinetics and remarkable performances of various solid catalysts. Initial AFCs were operated at 200 ~ 240℃ in KOH electrolyte in 1950s. Porous NiO catalysts were used as cathode materials with good performances. The three-phase boundaries could be well maintained, although no PTFE materials were available early in 1960s.

In general, AFCs outperform PAFCs in terms of efficiency. As Pt catalysts are used as cathode materials in a 30% KOH electrolyte, the output potential could be as high as 0.87 V. Higher efficiency means higher current density under high potentials, which enable AFCs as better candidates for space programs.

The most detrimental impurity for AFCs is CO_2. In late 1970s, a fuel processor has been successfully implemented prior to entrance of the cells using Pt/Pd catalysts. Similarly, soda scrubbing units can be implemented to minimize CO_2 concentration. Despite of such additional processing units, unfavorable cost and over-sized devices are still making such arrangement unpractical. Slow build-up of K_2CO_3 is still the main issues for portable and stationary applications.

The normal source for H_2 production is mainly the reforming of natural gas. The reformate consists of a mixture of H_2, CO_2 and other impurities such as CO, N_2 and unreacted CH_4.

Removal of CO_2 from the mixture is not difficult, but obtaining highly purified H_2 is still costive, to which the construction of AFCs can be very uneconomic.

During normal operations, the following aspects can be considered as advantages. (a) no drying-out occur for AFCs due to constant water content in caustic medium. (b) complicated heat exchanger compartments not necessary for solid electrolytes. (c) impurities such as carbonates can be resolved. (d) no detectable OH^- concentration gradient throughout the cell. (e) no build-up of gas bubbles between electrode and electrolyte.

The fundamental reactions for AFCs are discussed briefly here, with two half-cell reactions:
$$H_2 + 2OH^- \longrightarrow 2H_2O + 2e^- \text{ (Anode)}$$
$$O_2 + 2H_2O + 2e^- \longrightarrow 2OH^- \text{ (Cathode)}$$

Hydroxyl ions are conducting species in electrolyte. The overall reaction is still the H_2 reacting with O_2 to form water, electricity and heat (Figure 4 – 28).

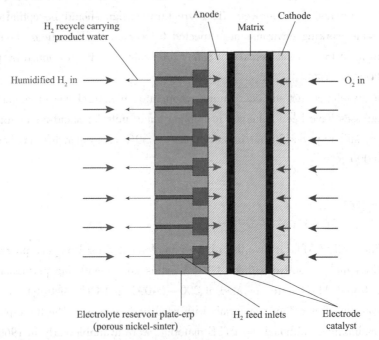

Figure 4 – 28 Ion transport in AFC

4.6.1 Cell compartments

Typically, KOH is immobilized in solid electrolyte matrix in application of space programs with 35% ~ 50% (wt) KOH for low temperature operations (< 120℃). Highly concentrated (85%) KOH is often immobilized for high temperature operation (260℃). Ni, Ag, Pt, metal oxides, spinels can be used as electrocatalysts to boost the two half-cell reactions.

The cell device used in Apollo program is cylindrical modules with 110 kg, producing approximately a peak power of 1.42 kW at 27 ~ 31 V. This type of cell was operated at high temperature at a moderate pressure (0.4MPa). The typical performance was 0.85 V at 150 mA/

cm². The cell device used in Space Shuttle Orbiter was rectangular with a width of 38 cm and a length of 114 cm, weighing 118 kg, which produced a peak power of 12 kW at 27.5 V. This cell was operated at low temperature. For low temperature AFCs, noble PtPd, AgNi, AuPt and AgNi catalysts have been used in the presence of PTFE materials. A wide range of solid materials including K_2TiO_3, CeO_2, asbestos and ZrP gel have been used as solid matrix to immobilize KOH species. Gold coated Mg is used for bipolar plates. Raney Ni anodes with Raney Ag cathodes can also be used to achieve good performances.

H_2/O_2 AFCs using immobilized electrolytes has been fully developed at present. The key contributor for cost reduction is the life of cell materials, such as stack and frames. Corrosion of frames in AFCs is still the main issue, leading to a total stack life of < 3000 h, which is insufficient for stationary applications. Installing new stack components can enhance the life of AFCs in Orbiter up to 87000 h.

Other developments of cell materials include the multilayered electrodes. Porous electrodes durable in alkaline medium has been the research interests in academia. Instability of PTFE even at 200 ℃ is another concern which should be addressed properly. Other technological considerations include developing effective techniques to mix powers and carbon electrodes, sedimentation, spraying and sintering.

4.6.2 Performance

Early AFCs were operated under high temperature, as already mentioned. The performances of AFCs are shown in Figure 4-29. State-of-art AFCs are operated under ambient pressure and low temperature (<75 ℃). Pressures affect gas solubility over a wide range of temperature. The following equation can be applied to predict the voltage changes with varied operation pressure.

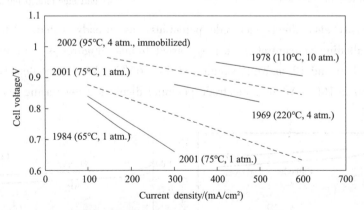

Figure 4-29 Performance curves for AFCs

$$\Delta V(mV) = 0.15 \times T(K) \times \lg \frac{P}{P_{ref}}$$

P_{ref} is the reference pressure at known potential. Elevated pressure is beneficial for faster kinetics. But operating pressures are often not higher than 0.5 MPa to avoid additional manufacture input and weight increase to sustain pressurized conditions.

Increasing temperature could enhance cell potentials, as kinetics are faster (Figure 4 – 30). Low temperature leads to lowered potential at given current density. High current density will cause more significant drop in potential at lower temperature. From the graph obtained from KOH AFC, it is found that, as temperature decreases from 90 ℃ to 60 ℃, the electrode potential can decrease as high as 0.5 mV/℃ at 150 mA/cm^2. The temperature coefficient as be in the range of 0.5 to 4.0 mV/℃.

Impurities in fuels lead to degradation of alkaline electrolyte. It will also increase the viscosity of electrolyte, resulting in poor electronic conductivity. Typically, CO_2 content higher than 350 ppm will cause significant drop in electrode potential. In addition, precipitation of carbonates on the porous electrode also block surface reactions. Figure 4 – 31 presents the voltage drop at electrode in the presence and absence of CO_2. Obviously, CO_2 contaminated air will result in short operation life from 3000 h to less than 1500 h.

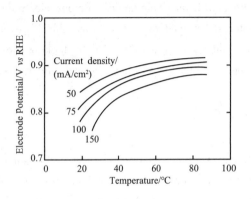
Figure 4 – 30 Influence of temperature on electrode potential

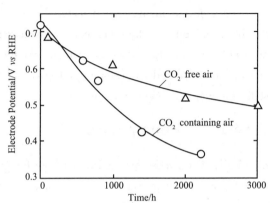
Figure 4 – 31 CO_2 contaminated air on cell electrode potential

Current density also affects electrode potentials, as already stated. Potential decreases under different alkaline concentration display varied trends. Figures 4 – 32 present electrode performances in 9 m and 12 m KOH solution at 55 ~ 65 ℃. As current density J (mA/cm^2) increases from 40 to 100 mA/cm^2, electrode potential displays a decreasing tend approximately of 0.2 ~ 0.5 ΔJ.

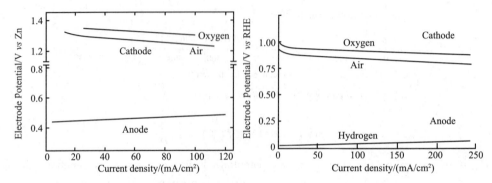
Figure 4 – 32 KOH concentration on cell electrode potential

4.7 PAFC

Phosphoric acid fuel cell (PAFC) was the first commercialized fuel cell technologies in the world. PAFCs have been widely used as stationary units for industrial and residential purposes. Most built units using PAFC technologies exceed the other types of fuel cells in terms of capacity ranges. Many plants are in 50 ~ 200kW.

Original PAFCs were operated with > 95% H_3PO_4 acid using supported Pt catalysts. CO_2 containing air was used as oxidant and H_2-rich gas were needed. The general scheme for PAFCs is shown in Figure 4 – 33.

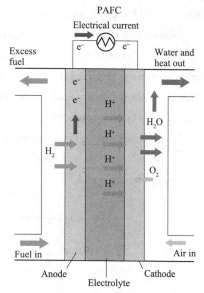

Figure 4 – 33 Schematic graph for PAFC

4.7.1 Cell compartments

Original PAFCs were manufactured with porous electrodes using PTFE bonded Pt black. In the past decades, carbon supported Pt catalysts have replaced Pt black in porous PTFE electrode structures. Pt loading has been reduced significantly from 9 mg/cm^2 to 0.1 mg/cm^2.

As for operating conditions, >200℃ with 100% phosphoric acid has been commonly used in PAFCs today. Pressures can be up to 0.8MPa in a 11 MW electric unit for demonstration plant. The following issues remains to be addressed.

Small scale PAFCs can be operated at atmospheric pressure although elevated pressure increase process efficiency. High pressure results in higher capital costs. Pressure may also increase the chance for corrosion. Phosphoric acid electrolyte (H_3PO_4) produces a vapor. This vapor, which forms over the electrolyte, is corrosive to cell locations other than the active cell area. Use of carbon black as structural materials can reduce the rate of corrosion, as it is relatively more stable than other alloy materials (Table 4 – 4). However, carbon corrosion and Pt leaching are still observed at cell voltage higher than 0.8 V.

Table 4 – 4 Components for PAFCs

Component	Since 1965	Since 1974	Current status
Anode	PTFE-Pt black	PTFE-Pt/C	PTFE-Pt/C
Cathode	PTFE-Pt black	PTFE-Pt/C	PTFE-Pt/C
Electrode support	Alloy mesh	Graphite	Graphite
Electrolyte support	Glass fiber	PTFE-Silicon carbide	PTFE-Silicon carbide
Electrolyte	85% H_3PO_4	95% H_3PO_4	100% H_3PO_4
Electrolyte reservoir		Graphite	Graphite

Graphite materials serve as a support for electro-catalyst layers. A typical graphite structure in PAFCs has an initial porosity of approximately 90%. The water-proof graphite structures contain 3 ~ 50 μm diameter to allow facile gas permeability. The composite structure consisting of a carbon black and PTFE layer for stable three-phase interface in the fuel cell.

Similar to PEMFCs, bipolar plates separate individual cell and electronically connect them in series in cell stack. Gas channels are obviously needed to allow feed gases to diffuse through porous materials and enable removal of products and inerts. The all-graphite bipolar plates are sufficiently corrosion-resistant for an estimated life of 40,000 hours in PAFCs, but they are still relatively costly to manufacture.

The atmospheric pressure short stack with 32 cells can be operated with an initial performance of 0.65 V/cell at 215 mA/cm^2. The performance degradation was lower than 4 mV/1,000 h operation (Figure 4-34).

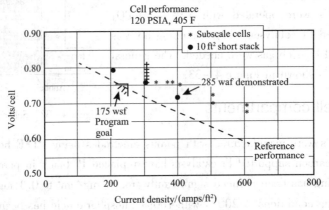

Figure 4-34 Water-cooled PAFC performances

For catalyst materials, Mitsubishi Electric Corporation studied alloyed catalysts with thinner layers to increase metal utilization efficiency. These improvements resulted in a better initial performance of 0.5 mV and 0.915 W/cm^2, although it is still a demonstration unit in small scale.

Improvement of catalyst performance and protection of carbon materials against corrosion are the main focus for technological progress. Organic molecules such as tetramethoxyphenylporphyrins (TMPP), phthalocyanines (PC), tetraazaannulenes (TAA) and tetraphenylporphyrins (TPP) have been studied to improve O_2 reduction rate as co-electro-catalyst materials. However, instability of organic molecules under harsh conditions in concentrated phosphoric acid. Interestingly, heat treatment of those molecules lead to formation of pyrolyzed residue, which display good activity and durability for O_2 reduction reaction.

Conventional cathode catalysts consist of nearly 10% (wt) Pt loading on carbon black support. Johnson-Matthey Technology Center has shown remarkable data for Pt-based alloyed catalysts with almost 40% enhancement for specific activity.

Contaminates in H_2 feed is also an issue for PAFCs. Current practice is to place sulfur cleanup systems and CO shift converters prior to the cell (normally in the fuel processor before reforming) to reduce the fuel stream contaminant levels to the desired amounts. Experimental data from demonstration plant has shown that, an anode catalyst, was identified that resulted in only a 24 mV loss when exposed to a 75 percent H_2, 1 percent CO, 24 percent CO_2, 80 ppm H_2S gas mixture at 190℃, 85 percent fuel utilization, and 200 mA/cm^2. A $7/kW increase resulted with the 5 percent CO gas (compared to a 1 percent CO gas) at a 50 MW size.

4.7.2 Performance

The operating time for PAFC stacks have now exceeded 40,000 h. Five factors strongly affect the life performances of PAFCs: (a) heat-treatment temperature of cathode materials, (b) operating temperatures, (c) gas crossover through a matrix layer, (d) fuel starvation, (e) impurities.

Similar to other fuel cells, the polarization of cathode is greater with air (560 mV at 300 mA/cm^2) than with pure oxygen (480 mV at 300 mA/cm^2) because of dilution of the reactant. Increasing cell pressure can improve performance of PAFCs according to the following equation:

$$\Delta V(\text{mV}) = 146 \times \lg \frac{P}{P_{\text{ref}}}$$

The experimental data has suggested that the above equation is a reasonable approximation for a temperature range of $177°C < T < 218°C$. Other measurements for the same increase in pressure from 0.4 ~ 0.9 MPa, but at temperatures above 210°C show less agreement between the experimental data.

Temperature ranges affect intrinsic kinetics of Pt catalysts (Figure 4 − 35). Typically, the potential enhancement at elevated temperature can be estimated by the following equation:

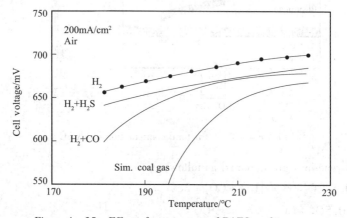

Figure 4 − 35 Effect of temperature of PAFC performances

$$\Delta V(\text{mV}) = 1.15 \times (T_2 - T_1)$$

The use of air with ~ 21% O_2 instead of pure O_2 results in a decrease in the current density of about a factor of three at constant electrode potential (Figure 4 − 36). Higher O_2 utilization efficiency also results in stronger polarization on the surface of electrode. At a nominal O_2 utilization of 50% for prototype PAFC power plants, the additional polarization estimated is 19 mV.

Polarization at Cathode (0.52 mg Pt/cm^2) as a Function of O_2 Utilization, which is

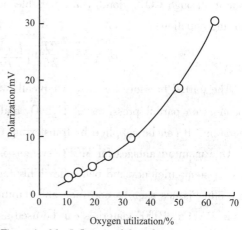

Figure 4 − 36 Influence of O_2 utilization on overpotential

Increased by Decreasing the Flow Rate of the Oxidant at Atmospheric Pressure 100% H_3PO_4, 191℃, 300 mA/cm², 0.1MPa.

4.8 MCFC

Molten carbonate fuel cells (MCFCs, Figure 4-37) were developed for natural gas and coal-based power plants for industrial purposes, and military applications. MCFCs are typically operated above 650℃. One advantage of high temperature operation is that noble metal catalysts are not required for electrochemical oxidation and reduction reactions.

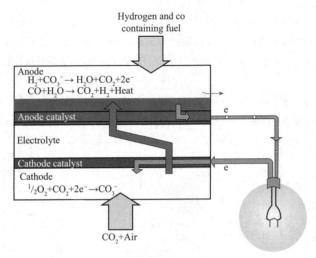

Figure 4-37 Structural configuration of MCFC

The half cell reactions are depicted as follow:

At anode: $H_2 + CO_3^{2-} \longrightarrow H_2O + CO_2 + 2e^-$

At cathode: $0.5O_2 + CO_2 + 2e^- \longrightarrow CO_3^{2-}$

According to the half-cell reactions, it is found that CO_2 need to transfer from cathode to anode to through CO_3^{2-} ion. The reversible potential for MCFC can be calculated based on the following equation:

$$E = E^o + \frac{RT}{2F}\ln\left(\frac{P_{H_2} P_{O_2}^{0.5}}{P_{H_2O}}\right) + \frac{RT}{2F}\ln\left(\frac{P_{CO_2,cathode}}{P_{CO_2,anode}}\right)$$

The partial pressure for CO_2 is typically varied in real cells. The cell potential is obviously dependent on partial pressures of H_2, O_2 and water. CO_2 is needed at cathode for CO_3^{2-} ion generation. It can be supplied by transferring CO_2 from anode exit to cathode inlet gas.

One main advantages for MCFC systems is high operating temperature. It leads to relatively higher system efficiency and better flexibility for various fuels. But elevated temperatures obviously cause other issues such as corrosion and stability of electrolyte under such condition.

In PAFCs, PTFE materials can be used as structural binder and wet-proof to maintain device integrity and ensure durable performances of electrodes and electrolytes. The phosphoric acid can be retained in the matrix of silicon carbide porous structures. However, PTFE cannot sustain high temperature operation in MCFCs. At present, suitable materials for MCFCs to establish good structural integrity under this condition. An alternative approach is often employed. As illustrated in Figure 4-38, The stable gas-electrode interface in MCFCs' porous electrodes are maintained through a balance in capillary pressures. By properly tuning the pore size in the electrode with electrolyte matrix, the optimal electrolyte distribution in matrix can be established. The electrolyte matrix can be filled with molten carbonate, while porous electrodes are partially filled.

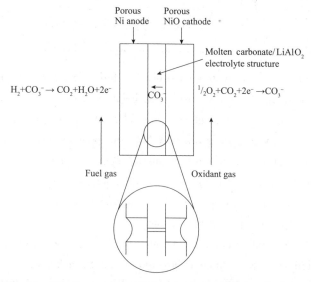

Figure 4-38 Dynamic equilibrium in porous MCFC cell elements

The extent of filling is determined by the equilibrium pore size in the compartment. If the pore size is smaller than the equilibrium pore size, the pore will be filled with electrolyte, otherwise the pore will be empty. Optimum distribution of molten carbonte electrolyte in different cell compartment is critical for high performances of MCFCs. Corrosion reactions, potential-driven migration and creepage of salt in the cell can contribute to redistribution of molten carbonates thus affecting the cell performances.

4.8.1 Cell compartments

Cell components in MCFCs have gone through evolution for electrode and electrolyte materials. Table 4-5 presents the evolution of those materials since 1965. Early years of MCFCs used precise metals for electrode such as Pt, Pd, Ag. Soon the technology shifted to use Ni-based alloys at anode and cathode. Since 1970s, molten carbonate/$LiAlO_2$ has been used and almost unchanged through the years' development. A major breakthrough in structural update is manufacture of electrode materials. Performances for electrode and endurance for stacks of MCFC have been dramatically improved, from 10 mW/cm^2 to 150 mW/cm^2. Another breakthrough occurred in 1980s, when endurance of MCFC stacks had dramatically improved. Cell areas could be built up to 1 m^2.

Table 4-5 Changes of cell component for MCFCs since 1975

Component	Since 1965	Since 1975	Current status
Anode	Pt, Pd, Ni	NiCr	NiCr, NiAl, NiAlCr 45%~70% porosity 0.2~0.5 mm thickness
Cathode	Ag_2O or NiO	NiO	lithiated NiO-MgO 70%~80% porosity 0.5~1 mm thickness
Electrolyte	Li-Na	Li-K	Li-K, Li-Na
Electrolyte support	MgO	Mixture of α-, β-, and γ-$LiAlO_2$	γ-$LiAlO_2$ 0.1~12 m^2/g

Conventional processes for fabricating electrolyte materials involved hot pressing of $LiAlO_2$ (> 5000 psi) and alkali carbonates (50% (vt)) just below melting temperature of carbonate salts (~490℃). The resultant electrolyte materials are relatively thick (1~2 mm) and difficult to produce in large scale. The void spaces are only 5% and poor uniformity of microstructures may cause uneven distribution of electronic current thus hot spots occur to shorten enduring life of cell stacks and high ohmic resistance. Alternative techniques such as tape casting has been reported. It also involves dispersing ceramic powder in a solvent which contains dissolved binders, plasticizers, additives to product proper slip rheology.

The composition of electrolyte affects endurance of MCFCs in different aspects. Higher ionic conductivities and relatively lower ohmic polarization, can be achieved with Li-rich electrolytes owing to high ionic conductivity of Li_2CO_3 compared to Na_2CO_3 and K_2CO_3. However, gas solubility and diffusivity are lower, with rapid and more severe corrosion in Li_2CO_3.

Mechanic robustness is a major consideration for NiO cathode materials. Because dissolution of NiO species under harsh temperature may cause unwanted issues thus leading to performance decay by redistribution of electrolyte in the cell. Sintering and mechanical deformation under compressive load induces dissolution of NiO in molten carbonate electrolyte, despite low intrinsic solubility of NiO in carbonates (<10 ppm).

If Ni ions diffuse in the electrolyte towards anode, and meet H_2, they could be reduced to metallic Ni species, which act as a sink for Ni ions thus facilitating dissolution of NiO from cathode. This is because under high CO_2 pressure, NiO can react and form carbonate.

$$NiO + CO_2 \longrightarrow Ni^{2+} + CO_3^{2-}$$

Anode materials. Ni—Cr alloy anodes are subjected to creep when placed under the torque load required in the stack to minimize contact resistance between components. The presence of Cr, although eliminating anode sintering, may consume carbonate during operation. Thus Ni—Al alloy materials with low Cr content are favorable for modern MCFC structural materials. The low creep rate with this alloy is attributed to the formation of $LiAlO_2$ dispersed in Ni substrate.

Needs for increasing sulfur tolerance are strong for anode materials. Anode materials can be doped with Mn and Nb species to enhance poisoning resistance. This is particularly important for converting coal-based fuels. However, limited progress has been made in this area.

Cathode materials. State of art cathode materials are mainly made of lithiated NiO. Dissolution of cathode materials has been considered as the primary life-limiting constraint for MCFCs. Increasing basicity of electrolyte can be a good option to reduce dissolution. Li/NaCO$_3$ composites are thus employed. Lowering CO$_2$ partial pressure can also assist alleviating the problem. LiFeO$_2$ cathodes also show good stability. But such type of electrode perform poorly compared with NiO materials. Co-doped LiFeO2 and Li-doped NiO provide promising performances.

Electrolyte matrix. Electrolyte structures are typically comprised of tightly packed fine α-or γ-LiAlO$_2$ with fiber or particulate reinforcement. Long-term measurement reveals that significant particle growth and phase transformation from γ to α leading to irreversible damage to pore matrix. Particle size grows rapidly under elevated temperature in the presence of CO$_2$. Such particle growth and phase transformation are induced by dissolution-precipitation mechanism. The electrolyte matrix should be robust enough to sustain operating mechanical and thermal stress but still maintain gas seal. Thermal cycle can also induce cracking due to unevenly distributed stress. In this context, ceramic fibers are most effective against cracking. Considering the advantages of ceramic materials, additional efforts should be put into improving the mechanical, conductive and operating properties.

Electrolyte. Li$_2$CO$_3$/K$_2$CO$_3$ (62 : 38 mol%) and LiCO$_3$/NaCO$_3$ (52 : 48 to 60 : 40 mol%) are good for various operating conditions. As already stated, electrolyte composition affects electrochemical activity, corrosion and decay rates. Evaporation of the electrolyte is a life-limiting factor for MCFCs. To this regard, Li/Na based electrolyte is better for high-pressure operation. Thicker electrolytes lead to prolonged operation time by preventing internal precipitation.

Bipolar plate. Current bipolar plates include a separator, current collectors, and wet seal. Separator and current collectors are Ni-coated 316L and wet seal is formed by aluminization of metal. The plate is actually exposed to anode and cathode environment. Low O$_2$ pressure prevents formation of non-conductive thin film. But reactions with creeping electrolyte, heat resistant alloys form thin corrosion layers. More expensive nickel-based alloys resist corrosion. Thermodynamic stable nickel coating is needed to protect the anode side.

4.8.2 Performance

MCFCs are usually operated at atmospheric pressure or under a slight overpressure (~0.4MPa). Carbonate ions are only providing ionic conduction in the temperature of 550 + ℃. The conversion of the electrochemical reactions is never complete owing to concentration polarization effect. Typically, H$_2$ conversion is in the range 65% ~ 80% with CO$_2$ conversion in the range 50% ~ 60%. O$_2$ is normally in large excess. MCFCs can supply current densities in the range 0 ~ 1500 A/m^2 and cell voltages in the range 0.6 ~ 1 V as a function of the operating conditions.

Increasing operating pressure of MCFCs lead to higher cell voltage owing to enhancement of partial pressure of reactants. But disadvantage of increasing pressure may cause Boudouard reaction (carbon deposition) and methane formation (CO hydrogenation).

Figure 4 - 39 illustrates the CO$_2$ effect on voltage enhancement. An empirical correlation for voltage changes from a pressure increase can be expressed as follow:

$$\Delta V(mV) = (76 \sim 84) \times \lg \frac{P_2}{P_1}$$

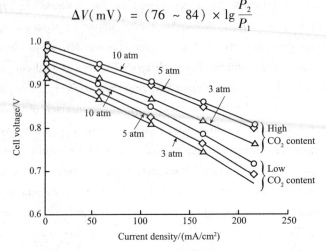

Figure 4-39   Influence of cell pressure on the performance of a 70.5 cm² MCFC at 650℃

Temperatures affect cell performances by altering kinetic rate and shifting equilibrium of reaction. For example, the equilibrium constant K is defined as:

$$K = \frac{P_{CO} P_{H_2O}}{P_{H_2} P_{CO_2}}$$

Temperature changes lead to compositional variation thus output potential can be altered accordingly. Some empirical equations have been summarized to correlate pressure increase with increasing temperatures.

$$\Delta V(mV) = 2.16 \times (T_2 - T_1) \quad (575℃ < T < 600℃)$$
$$\Delta V(mV) = 1.40 \times (T_2 - T_1) \quad (600℃ < T < 650℃)$$
$$\Delta V(mV) = 0.25 \times (T_2 - T_1) \quad (650℃ < T < 700℃)$$

Actually operating temperature beyond 650℃ may have detrimental effect on structural materials and loss of electrolyte thus cell performances might be compromised. Since MCFCs are often operated to consume fuels from natural gas and coal conversion, one should consider the possible detrimental effect caused by impurities in the gas feeds. A list of possible consequence of introducing impurities has been presented in Table 4-6.

Table 4-6   Type and effect of contaminants in MCFCs

Type	Contaminant	Effect
Particulates	Coal fines, ash	Plugging gas passages
Sulfur	H_2S, organic sulfurs	Voltage loss Reaction with electrolyte
Halides	HCl, HF, HBr	Corrosion Reaction with electrolyte
N compounds	NH_3, HCN, N_2	Reaction with electrolyte
Metals	As, Pb, Hg, Cd, Sn, etc.	Deposition on electrode Reaction with electrolyte
Hydrocarbons	C_6, C_{10}, C_{14}, etc.	Carbon deposition

Current density affects voltage by changing magnitude of ohomic loss. Higher current density tends to cause concentration polarization.

Internal reforming is incorporated with MCFCs to obtain additional H_2 in fuel feed. Internal reforming is combined with electrochemical reactions to close the thermal loop. Therefore, two alternative approaches for internal reforming options, indirect and direct reforming (Figure 4 – 40). Indirect reforming is separated and adjacent to fuel cell anode. This type of cell can take advantages of exothermic electrochemical reactions and endothermic reforming reactions. Direct internal reforming does not have a direct physical effect on heat use. But CH_4 reforming cannot be effectively promoted to generate H_2 in such configuration.

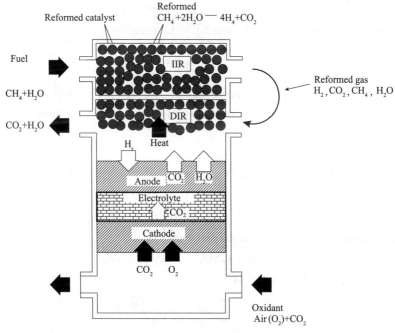

Figure 4 – 40 Indirect and direct reforming

4.9 SOFC

Solid oxide fuel cells (SOFCs) have an electrolyte consisting of a solid and non-porous metal oxide, often Y_2O_3-ZrO_2. The cell operates at 600 ~ 1000°C where ionic conduction by O^{2-} takes place. Such high temperature for SOFC poses challenges on structural materials. Low cost materials and cost-effective manufacture of ceramic structures are key contributing factors facing SOFCs (Figure 4 – 41). SOFCs allow conversion of a wide range of fuels, including various hydrocarbon fuels. Indeed, both simple-cycle and hybrid SOFC systems have demonstrated among the highest efficiencies of any

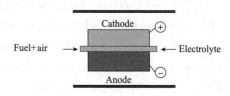

Figure 4 – 41 Schematic description of a SOFC

power generation system.

Reducing operating temperature to 700 ~ 800℃ could help lower the cost for construction materials. This would improve the economy of applications ranging from small-scale stationary power (~2 kW) to auxiliary power units for vehicles and mobile generators for civilian as well as military applications.

The basic reactions for SOFCs involve methane-air reactions.

$$CH_4 + 2O_2 \longrightarrow CO_2 + 2H_2O$$

Anode:
$$CH_4 + 4O^{2-} \longrightarrow CO_2 + 2H_2O + 8e^-$$

Cathode:
$$4 \times (0.5\,O_2 + 2e^- \longrightarrow O^{2-})$$

The cell voltage can be calculated by the following equation:

$$E = E^o + \frac{RT}{8F}\ln\left(\frac{P_{CH_4} P_{O_2}^2}{P_{CO_2} P_{H_2O}^2}\right)$$

In real SOFCs, the cell cannot promote directly the electrochemical conversion of fuels. Partial oxidation of fuel can occur to generate CO and H_2. Thus electrochemical reactions occur through conversion of syngas.

Anode:
$$2H_2 + 2O^{2-} \longrightarrow 2H_2O + 4e^- \quad CO + O^{2-} \longrightarrow CO_2 + 2e^-$$

Cathode:
$$1.5\,O_2 + 6e^- \longrightarrow 3O^{2-}$$

Thus overall reaction is expressed as the following:

$$2H_2 + CO + 1.5O_2 \longrightarrow CO_2 + 2H_2O$$

The Nernst equation for this system is,

$$E = E^o + \frac{RT}{6F}\ln\left(\frac{P_{H_2}^2 P_{CO} P_{O_2}^{1.5}}{P_{CO_2} P_{H_2O}^2}\right)$$

The Faraday efficiency is only 75% as the production of syngas is not an electrochemical reaction. In addition to this intrinsically reduced efficiency, current SOFC systems show very low fuel utilization (1%~8%) and process efficiencies.

4.9.1 Cell compartments

As already mentioned, electrolyte is made from ceramics such as Y_2O_3-ZrO_2 materials. Typically, the anode is a Ni-ZrO_2 cermet and the cathode is Sr-doped $LaMnO_3$.

Ni-based materials show good activity and selectivity as anodes in SOFCs. Composite anodes of Ni and samarium, gadolinium, or cerium oxides are being investigated by researchers. Despite good performances, structural degradation and volatilization are critical issues for Ni based anodes. Loss of Ni through hydroxide species in O_2-rich mixtures. Ni ions undergo redox cycles and easily form metallic Ni which precipitate on the surface of electrodes, with volumetric effect due to variation of oxidation states between NiO and Ni. It ultimately leads to damage of anode structures. O_2 react with Ni and form NiO, which lowers the oxidation rates of CH_4, which in turn reacts with NiO endothermically thus could possibly reduce cell temperatures.

Lanthanum strontium manganite (LSM) materials are most commonly used as cathode for SOFCs. Sintering temperatures affect the performances of LSM materials for O_2 reduction reaction. Doping LSM with other metal cations could alleviate the effect of sintering conditions. $La_{0.2}Sr_{0.8}Co_{0.8}Fe_{0.2}$ and $Ln_{0.7}Sr_{0.3}Fe_{0.8}Co_{0.2}$ are typical examples for modified LSM materials.

4.9.2 Cell design

From the geometrical point of view, SOFCs with coplanar electrodes consist of anode, cathode aligned side-by-side on the same side of electrolyte substrate. In the fuel-air mixture, a pressure difference for O_2 is established between the two adjacent electrodes (Figure 4-42).

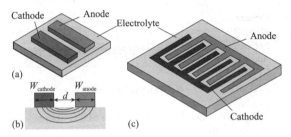

Figure 4-42 Cell design for electrodes

The following structural parameters will affect cell performances, electrode shape, inter-electrode gas, electrode width, thickness, electrolyte thickness, stacks, flow directions, gas intermixing, chemical interaction between coplanar electrodes.

Three Types of Tubular SOFC: (a) Conduction around the Tube (e. g. Siemens Westinghouse and Toto), (b) Conduction along the Tube (e. g. Acumentrics), (c) Segmented in Series (e. g. Mitsubishi Heavy Industries, Rolls Royce).

Tubular SOFC is by far the most developed model. Figure 4-43 illustrates the design for tubular pattern. The current is conducted tangentially around the tube. Each tube contains one cell. Tubes are connected either in series or in parallel. In micro-tubular scenario, current is conducted axially along the tube. Interconnections are made at the end of each tube. A metallic current collector made of Ag is often applied to minimize the in-plane resistance on the cathode side. Cathode tube is fabricated by extrusion and sintering. It has a porosity approximately 30%~40% along rapid gas diffusion to cathode/electrolyte interface. The interconnects doped with lanthanum chromite must be impervious to fuel and oxidant gases but with good conductivity. The material should be chemical stable under 1000℃ in the presence of oxidative environment.

A schematic description of gas manifold and cross section of a tubular bundle is presented in Figure 4-44. A major advantage for such design is the relatively large single tube cell where successive layers can be deposited without chemical interference with previously deposited layers.

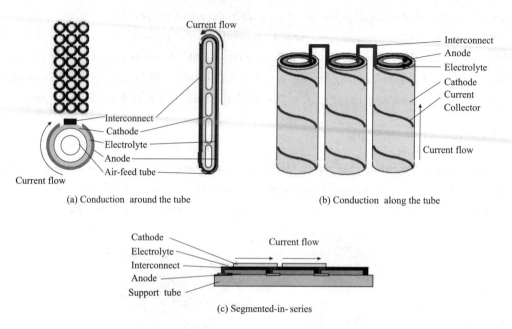

Figure 4-43 Schematic tubular SOFC design

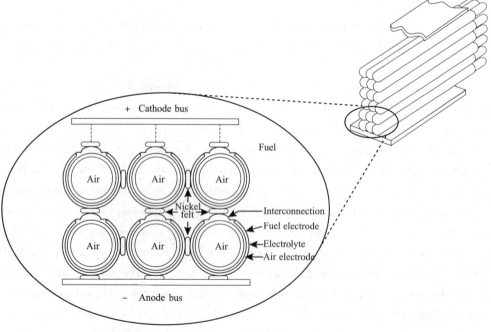

Figure 4-44 An example of tubular bundle

4.9.3 Performances

The output voltage for H_2-based SOFCs involves two half-cell reactions, shown as follow:

$$H_2 + O^{2-} \longrightarrow H_2O + 2e^-$$
$$0.5O_2 + 2e^- \longrightarrow O^{2-}$$

The corresponding Nernst equation can be expressed:

$$E = E^\circ + \frac{RT}{2F}\ln\left(\frac{P_{H_2}P_{O_2}^{1.5}}{P_{H_2O}}\right)$$

The thermodynamic efficiency for SOFCs on H_2/O_2 is lower than that of MOFCs and PAFCs, owing to lower free energy at elevated temperature. On the other hand, higher operating temperature might reduce the significance of polarization resistance.

Voltage loss in SOFCs is mostly determined by ohmic resistance in the cell. Cathode resistance is a major contributor for ohmic loss, despite of remarkable specific resistivities of electrolyte and cell interconnection because of short conduction path through those components but long current path in the plane of cathode materials (Figure 4 - 45). Flattened tube could be a good design to improve power density.

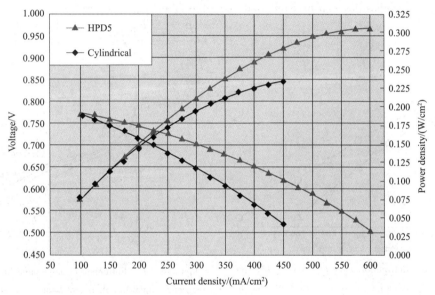

Figure 4 - 45 Performance curves for SOFCs with different geometries

The influence of pressure, temperature is similar to that of MCFCs and PAFCs. Enhanced pressure often leads to better output voltage. For example, an approximation equation for SOFCs operated under 1000℃ is shown as below:

$$\Delta V(\text{mV}) = 59 \times \lg\frac{P_{\text{cell},1}}{P_{\text{cell},2}}$$

Other correlation equations for temperatures and pressures have already been summarized in "Fuel Cell Handbook (7th edition, U. S. DOE, NETL report)". The influence of reactant composition and gradient will be discussed in details in this section.

(1) Gas composition and utilization

Since SOFCs are often incorporated with internal reformer using hydrocarbon as fuel feeds, effect of varied gas composition on output performances has been experimental studied extensively.

Another important issue is recycling of CO_2 from the spent streams, as SOFCs only utilize O_2 in the cathode.

Oxidant. Figure 4 – 46 present the influence of using air or pure O_2 on cell voltage in SOFCs. As expected, cell voltage at 1000 ℃ displays an improvement with pure O_2 as oxidant rather than with air. The difference in cell voltage with pure O_2 and air suggests that concentration polarization plays a critical role during reduction of oxidant. With increasing current density, such difference becomes more significant.

Fuel. The fuel feed composition could possibly vary from pure H_2, to hydrocarbons with negligible H_2. Figure 4 – 47 presents the open circle potential for SOFCs, where the following observations have been made according to experimental studies: (a) presence of H_2 in the feed results in higher potential, (b) higher O/C ratio leading to complete oxidation favors enhanced cell potentials. But high temperature is not favorable for high potential, as water gas shift reaction is equilibrium unfavorable at 1,000 ℃ compared with 800 ℃. The theoretical gain in voltage due to changes of H_2 composition can be expressed using the following equation:

$$\Delta V_{\text{anode}}(\text{mV}) = 126 \times \lg \frac{P_{H_2,\text{cell},1} \; P_{H_2O,\text{cell},1}}{P_{H_2,\text{cell},2} \; P_{H_2O,\text{cell},2}}$$

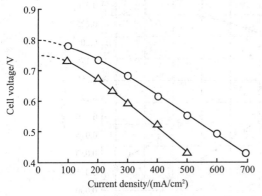 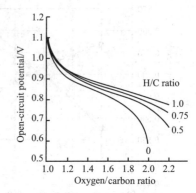

Figure 4 – 46 Cell performance at 1,000 ℃ with pure oxygen (○) and air (△) both at 25% utilization

Figure 4 – 47 Influence of gas composition of the theoretical open-circuit potential of SOFC at 1,000 ℃

(2) Impurities

Hydrogen sulfide (H_2S), hydrogen chloride (HCl) and ammonia (NH_3) are impurities typically found in coal gas. Early studies have simulated coal-based fuel feed with NH_3, HCl and H_2S as impurities. Experiments results showed almost no degradation in the presence of 5,000 ppm NH_3 in the feed. An impurity level of 1 ppm HCl also showed negligible degradation. But trace amounts of H_2S (~1 ppm) causes immediate performance decay.

(3) Cell life

Demonstration units for SOFCs have shown a 69,000 h operation with 0.5% voltage decay per 1000 h. This tubular design is based on the early calcia-stabilized zirconia porous support tube. In the current technology, the PST has been eliminated and replaced by a doped lanthanum manganite air electrode tube. These air electrode-supported (AES) cells have shown a power density increase of approximately 33% over the previous design.

4.10 Development of modern electrochemistry

4.10.1 Electrochemical measurement

(1) Electric systems

We have already introduced electrochemical system in solar energy chapter. In fuel cell testing and electric measurement experiments, two-electrode and three-electrode systems are both popular. The reduction reactions occur at cathode while oxidation reactions are carrying out at anode. Similarly, working electrode (WE) is defined the one with testing sample for target reaction. The counter electrode (CE) is an auxiliary electrode which allows current flow, with reference electrode for potential reference.

The configuration of a two-electrode system is illustrated as follow (Figure 4-48). The cell voltage, as ready defined, is the sum of theoretical potential at open circuit, change of potential at CE, voltage drop and change of potential at WE.

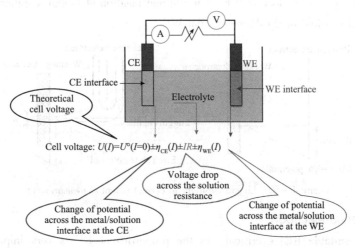

Figure 4-48 Two electrode system and cell voltage

$$V_{cell} = V^\circ \pm \eta_{CE} - iR \pm \eta_{WE}$$

It is important to mention that, the change of potential at electrode is actually contributed from activation and concentration overpotential. Therefore, this equation is similar to the following, defined in output voltage section.

$$V_{cell} = V_{cathode} - V_{anode} - iR - \eta_{con}$$

However, the above-shown configuration cannot independently measure any change in potential at WE and caused by electrolyte. Therefore, it is essential to minimize the change at CE. To achieve this purpose, the surface of CE is often designed to be much higher than that of WE. The introduction of reference electrode (RE) can correct such error caused by change of CE when focusing on the performances of WE (Figure 4-49). RE does not carry current. Since change of CE is already minimized, the change of WE can be measured as difference between WE and RE.

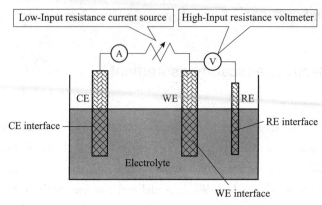

Figure 4-49　Three electrode system

Placing RE very close to WE could minimize the contribution from electric resistance to large extent. For this purpose, Luggin capillary is often used to create additional ionic path from RE to WE, to avoid any contamination of electrolyte with Cl^- and SO_4^{2-} as anions. As shown in Figure 4-50, the capillary is very close to WE. The optimal position of Luggin capillary is that, the tip is at a distance of two-fold of tip diameter from WE.

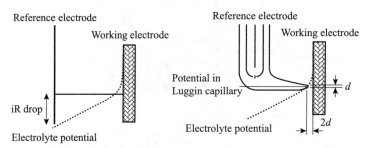

Figure 4-50　Luggin capillary to correct potential measurement

(2) Reference electrode

Selection of suitable RE is critical, as the properties of RE is very important to convert measured values into the actual performances of WE. The observed performances of WE highly depend on the nature of RE. In addition, appropriate comparison with literature data where another type of RE is often used, still remains a challenge. Some common REs are listed in this section.

Ag/AgCl (SSCE):
$$AgCl_{(s)} + e^- \Longrightarrow Ag_{(s)} + Cl^-_{(aq)}$$

Hg/Hg$_2$Cl$_2$ (SCE):
$$Hg_2Cl_{2(s)} + 2e^- \Longrightarrow Hg_{(liq)} + 2Cl^-_{(aq)}$$

Hg/HgO (MMO):
$$HgO_{(s)} + H_2O_{(liq)} + 2e^- \Longrightarrow Hg_{(liq)} + 2OH^-_{(aq)}$$

Hg/Hg$_2$SO$_4$(MSE):

$$Hg_2SO_{4(s)} + 2e^- \rightleftharpoons 2Hg_{(liq)} + SO_{4(aq)}^{2-}$$

Ag/Ag$_2$SO$_4$(SSSE):

$$Ag_2SO_{4(s)} + 2e^- \rightleftharpoons 2Ag_{(liq)} + SO_{4(aq)}^{2-}$$

Pt/H$_2$/H$^+$(RHE):

$$2H_{(aq)}^+ + 2e^- \rightleftharpoons H_{2(g)}$$

Finding a suitable RE is very difficult. The following properties need to be considered: (a) have a stable potential, (b) sustain charge transfer imposed by the instrument without changing its potential, (c) return to fixed reference potential after disturbance, (d) follow Nernst equation, (e) solid component sparingly soluble in electrolyte. This is the reason that most experimental studies have employed Ag/AgCl and Hg/Hg$_2$Cl$_2$ types of RE. Ag/AgCl REs have been widely used owing to simplicity, quietness and robustness. However, if the cell does not have Luggin capillary tip, which isolates RE from the main cell, the two REs might not be the best choices. In some cases, trace amounts of ions such as Cl-may adsorb on the WE and dramatically modify the electrochemical properties. Under such condition, Hg/Hg$_2$SO$_4$ RE is a good choice.

The filling solution in RE might rise if the cell is operated in alkaline electrolyte. Therefore, Hg/HgO and Pt/H$_2$/H$^+$ REs are considered as good candidates. Pt/H$_2$/H$^+$ RE has the advantage of filling the same type of solution for stable and reliable reference potential. The measured properties against other REs can be normalized using the following equations.

$$E_{RHE} = E_{SSCE} + 0.059pH + E^\ominus_{(SSCE)}$$
$$E_{RHE} = E_{MMO} + 0.059pH + E^\ominus_{(MMO)}$$
$$E_{RHE} = E_{SCE} + 0.059pH + E^\ominus_{(SCE)}$$

(3) Cyclic voltammetry (CV)

The fundamentals involved in cyclic voltammetry are applying a linear variation of the potential between an initial value and a final value to the WE, reproducing the variation periodically as long as necessary. The electrochemical response as current. We can obtain current as function of potential namely as potential-intensity curve (Figure 4-51). The current response is often referred as **voltammogram**. It will provide signature information for a material in a given WE. To eliminate or minimize the

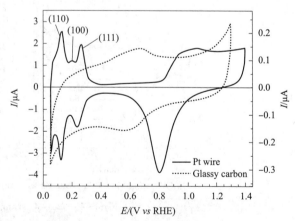

Figure 4-51. Example of voltammogram for Pt and carbon

effect caused by migration of undesired ions and provide sufficient ionic conductivity, a concentration of higher than 0.1 M of electrolyte is favorable. The voltammogrames provide fundamental and qualitative information on the nature of the reversibility of redox reactions in fuel cells. These features make CV as most appropriate method to study electrochemical reactions.

The cell should be cleaned to eliminate the presence of impurities, prior to experiments, as trace amounts will affect the CV curve. Pt wires are often used to probe the cleanness of the electrochemical cell. Three characteristic peaks for crystallographic planes (110), (100) and (111) will appear associated with H_2 desorption, at 0.13, 0.20 and 0.26 V on RHE.

Typical CV curves of Vulcan carbon supported Pt and Pd catalysts along with this technique are described to assess the electrochemically active surface area (ECSA). CO stripping is usually used to evaluated the durability of catalysts in the presence of organic compounds. This technique consists of adsorbing CO molecules dissolved in aqueous medium under controlled potential and oxidation of CO through a linear variation of electrode potential. Such measurement will evaluate the catalytic activity as CO is the smallest possible intermediate during electrooxidation of organic compounds. It can also probe the catalyst surface to evaluate ECSA.

More importantly, combining CV information with spectroelectrochemistry, basic information on adsorption mode including linear or bridge can be determined properly. This is because once CO adsorbs on the catalyst surface, it will affect irreversibly the metal atoms on the surface. Initially, CO adsorption is measured at given electrode potential under bubbling condition for a certain period of time, followed by daerating by inert gas for at least 30 min before stripping potential measurement (scan). CO stripping CVs are then collected on electrode. Several domains are observed, Dones I and II are characterized by formation of metal oxide (e.g., PdO in Figure 4-52). Domain III occur when CO adsorption is carried out long enough (>5 min), suggesting the monolayer of CO on catalyst surface. Clearly, different domains can be used to characterize the state of CO on catalyst surface.

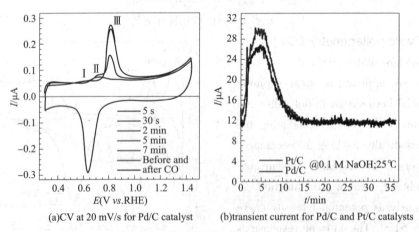

(a) CV at 20 mV/s for Pd/C catalyst   (b) transient current for Pd/C and Pt/C catalysts

Figure 4-52 CO stripping experiments

To be more specific, determination of ECSA can be illustrated through the following two cases. For Pt/C catalyst, adsorption and desorption of Pt-H species are used to characterize the active surface area due to the formation of monolayer of H. For Pd catalysts, since it is readily dissociating H_2 into atomic H in lattice, the reduction peak for PdO and CO stripping methods are valid for determining ECSA. In general, a charge density of 210 $\mu C/cm^2$ is associated with Pt-H monolayer, while 424 $\mu C/cm^2$ is associated with reduction of PdO monolayer. The regions of interest for Pt-H and PdO monolayers are presented in red in Figure 4−53. The exchange charge (Q_{ex}) can be quantitatively calculated by integration method or weighing method (Lavoisier's approach).

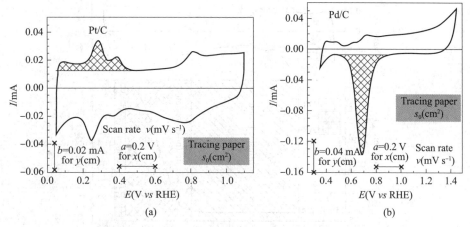

Figure 4−53 Typical CVs for (a) Pt/C and (b) Pd/C catalysts

The integration method is validated as exchange charge can be calculated using the following equation:

$$Q_{ex} = \frac{1}{v}\left(\int_{E_{onset}}^{E_{end}} i dE\right)$$

The physical meaning for integration is actually the shaded area shown in Figure 4−53. Alternatively, weighing method can also be employed. The philosophy is simple and straightforward. Cutting down the tracing paper and weigh the mass of shaded area over the whole area under the CV curve will give the exchange charge value.

(4) Kinetic activity

Catalytic electrochemical reactions occur at both anode and cathode. Electrochemical reactions are taking place on the surface of supported Pt catalysts. TEM images can be found in numerous publications. A typical CV curve for Pt/C catalyst is shown in Figure 4−54. Variation of potential is imposed on Pt/C electrode starting from 0.05 V (vs RHE) and the potential is scanned towards more positive values. Oxidation of Pt atoms occur when potential is higher than 0.7 V. PtO_x can then be reduced at approximately 0.75 V followed by adsorption of H^+ in acidic medium during negative potential scan (0.45~0.2 V).

Activation overpotential curves for ORR are divided into three types of regions. The graph is similar to what have been shown in previous section. Under high output potentials, ~1.05 V, ORR is limited by activation barriers. With increased current density, reaction become more kinetic controlled. Continuous enhancement in current density moves reaction into diffusion controlled regime. Kinetic regime is characterized by half-wave potential ($E_{1/2}$), which is half of the disk density. The reaction rate data can be collected for kinetic measurement to assess the activity of an electrocatalyst. More positive $E_{1/2}$ value suggests higher activity of the catalyst. Onset potential (E_{onset}) is defined as the potential where ORR current starts to appear. This value should be collected for current density approximately 1% of that under diffusion controlled regime. Half-wave and onset potentials are important parameters characterizing the performances of electrocatalysts.

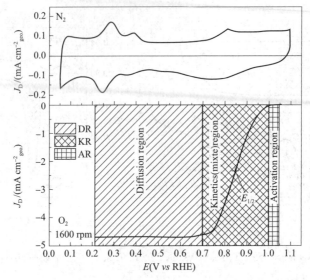

Figure 4-54 CV curve for Pt/C catalyst and overpotential analysis for different regions

4.11 H_2 as important energy carrier

4.11.1 Importance of H_2 for renewable energy

Hydrogen element is the most abundant one in the universe. It mainly exists as molecular H_2 with 100ppm present in our atmosphere. H_2 is colorless, odorless and non-toxic gas. It has a very high energy content of approximately 143 MJ/kg and combustion with O_2 only produce water as final product (Table 4-7). H_2 production need consuming primary energy such as natural gas, coal, solar or biomass. One important piece of fact is that, currently, approximately 96% of today's H_2 production utilizes fossil fuels emitting CO_2 in the process (Figure 4-55). Therefore, H_2 production from non-renewable source is actually not green and sustainable as it contributes to the accumulation of greenhouse gas emission in similar way as direct combustion of carbon-based feedstocks. It is generally accepted that, carbon-neutral H_2 production should be achieved through transformation of renewable resources such as biogas, solar energy, bio-ethanol, *etc*.

Table 4-7 Properties of H_2, gasoline and natural gas

	H_2	Gasoline	Natural gas
Explosive envelop/%	4.1~75	1.4~7.6	5.3~15
Heat duty/MJ	0.02	0.2	0.29
Diffusivity/($\times 10^{-5} m^2/s$)	6.11	0.55	1.61
Energy density/(MJ/kg)	143	44	42

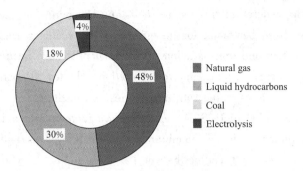

Figure 4-55 Feedstocks and technology for H_2 production

As already mentioned in previous sections, the quality of H_2 (purity) is extremely important for durable and efficient operation of fuel cells, particularly for low temperature devices. Today, H_2 is mainly produced through thermochemical processes, despite extensive efforts devoted in photochemical and electrochemical water splitting technological research. Steam reforming, autothermal reforming, partial oxidation, pyrolysis and gasification processes are often employed at industrial level. In most of those conversion processes, generation of syngas consisting of CO, H_2, CO_2, H_2O is often involved followed by shifting reaction making H_2-rich gas and purification step.

Electrochemical water splitting has becoming increasingly popular research topic in recent decades, as it utilizes solar, electrical and thermal energy to split water molecules directly to H_2 and O_2 products. H_2 with good purity for specific requirement can be produced in a more energy efficient way with reduced requirement for post-treatment steps. Electrical energy is primarily converted to H_2 through water electrolysis. Solar energy can be used to induce electrons and enable water splitting reaction.

4.11.2 Global Efforts and Advances on H_2 as Energy Source

Several international organizations such as International Energy Agency (IEA) have forecasted that, the world will be emerging into decarbonized H_2 economy. Undoubtedly, fuel cells will become the future dominant energy conversion technology in transportation, distributed energy systems, as well as other portable applications. H_2 represents one of the main fuel options in future energy industry for sustainable growth of global economy. Since Japan published its "Basic Hydrogen Strategy" in 2017, several other governments, including most recently Spain, Germany and the Netherlands, as well as the European Commission, have put forward H_2 strategies and roadmaps, and more are expected in coming months.

(1) H_2 energy in United States

H_2 energy has been one of the most important strategic energy sources recognized by the government of United States. The concept of "H_2 economy" has been proposed since 1970s. Bush administration and Obama administration proposed the blue map and global energy strategies, respectively to accelerate relevant research efforts in commercializing H_2 technologies. U.S. government selected October 8th as national fuel cell day since 2018.

In the past decade, federal funding for fuel cells and H_2 energy has exceeded $1.6 billion. Tax reduction has also been beneficial for the growth of H_2 economy with vast construction of infrastructures for H_2 station and transportation. Additional bills have been approved by congress on investing stationary energy systems and transportation vehicles. Department of Energy also established research teams collaborating with universities and institutes.

The number of patent filed in U.S. is second highest in the world following Japan. Patents filed in the areas of proton exchange membranes, fuel cell systems and portable H_2 storage technologies show that U.S. is still leading the world in those fields. U.S. and Japan contribute to almost half of global patent applications in the world. The capacity of liquefied H_2 and transportation H_2 cars in U.S. is the highest in the world.

Till the end of 2018, there had been 42 H_2 station under construction, the number of which rises to 75 and 200 in 2020 and 2025, respectively. The stationary devices have exceeded 100 MW construction per year, with a total of 500 MW at present.

(2) H_2 energy in European Union

EU has been the worldwide leader in utilizing alternative energies such as solar, biomass and H_2. Regulations on climate changes and energy strategic plans passed by EU have proposed fuel cells as promising technologies for stationary and portable devices. Budget for H_2 station construction is approximately €665 million from 2014—2020. Approximately 152 H_2 stations have been under construction. This number will increase to 770 till 2025. About 142 fuel cell vehicles have been used in public transient systems in Italy, Britain, Germany, Demark, with a capacity of 28.8 MW.

Germany has strategized H_2 and fuel cells as future energy technologies. NOW-GmBH organization has promoted development of fuel cell technologies since 2006. A total of €1.4 billion has been invested from 2007 to 2016 and more than 240 companies have been funded to commercialize fuel cell technologies. Germany has a world No.1 capacity for H_2 production from renewables and ranks 3rd on manufacture industry for fuel cells. German companies have been devoting in developing Power to Gas technologies to produce additional H_2 from natural gas sources.

(3) H_2 energy in Japan

Japan is leading H_2 utilization and fuel cell technologies across the globe. In the past 30 years, Japanese government have invested hundreds of billion yen for fundamental and technological research in this field. The maximum output power for commercialized Mirai Toyota products is 114 kW, which can work at-30℃ and charging only for 3 min for ~400 millage. Over 7000 vehicles have been manufactured and sold, accounting for 70% of transportation fuel cell units. EneFarm has eastalibhsed 2,740,000 devices for family appliances, with a total reduction of cost by 69% since 2009. Japan has 113 commercial H_2 station in use with additional 320 to be infrastructed by 2025. A total of 800,000 fuel cell vehicles will be in use by 2030.

(4) Global efforts on "green H_2"

IEA has initiated extensive R&D activities around the globe to promote low carbon H_2

manufacture. A few reports and important information have been collected in this book and listed as follows:

(a) Carbon-free hydrogen from low-cost wind power, stored for use on demand. This project aims to demonstrate the production of H_2 for storage, deployment and grid balancing. Through performance testing, this project shows the potential for power-to-H_2 technologies while validating the innovative proton exchange membrane (PEM) electrolyzer and delivery processes.

This project has been practiced in Denmark, where 50% of electricity consumption derives from wind mills. The wind electricity is converted into chemical energy in an advanced electrolyzer where water splitting into H_2 and O_2 occurs. The as-produced H_2 is stored for later delivery (via pipelines and trailers) for use by customers for clean transport and industrial purposes.

The project was initiated in 2015 and the technical facility was completed in 2017. H_2 delivery began in 2018 and electrolyzer performance testing was concluded in September 2020 (facility shown in Figure 4-56). Key business results include the production and sale of H_2, provision of grid services since 2020, and real-world grid-interconnected use of PEM electrolyzer.

Figure 4-56 The facility in Hobro, Denmark

(b) Green refinery hydrogen for Europe. There is a strong incentive to reduce refinery CO_2 emission for large-scale H_2 production. Refineries are giant H_2 consumers, most of which is supplied by conversion of fossil fuels. The project named as "REFHYNE" will be installed and operated with a 10 MW electrolyzer in a large German refinery. Practically, the electrolyzer will provide bulk quantities of "green" H_2 produced without any greenhouse gas emissions. Such H_2 can be well integrated with traditional refineries, including desulfurization of fossil fuels.

Another function of REFHYNE project is that (Figure 4-57), it will also report on the conditions where electrolyzer business models will become more viable, in order to offer case-based data required to justify changes in existing policies.

Other values for REFHYNE project is, to test run the possible decarbonized industrial processes and increase local and domestic energy security. Obviously, it will also provide grid balancing options and create new local business and job opportunities.

(c) The growing momentum for hydrogen in Latin America. As another emerging economic region, Latin America could become a key contributor to the global efforts towards low-carbon H_2.

In line with global economic recovery since 2008, and recently COVID-19, governments in the region are taking decisive steps to foster potential of H_2 as a key component of their clean energy transitions and a source of export revenues.

Figure 4-57 Construction site of the REFHYNE project in Wesseling, Germany

Latin American Ministers and energy leaders stress need for sustainable and inclusive recovery from the COVID-19 crisis (Figure 4-58). At roundtable meeting hosted by IEA, Ministers and top officials from Latin American countries representing close to 70% of region's energy demand underscore opportunities to build back better. This meeting highlighted the need to keep up momentum behind the deployment of renewable technologies like wind and solar while ensuring security of supply, sufficient investments in grids and strengthening of regional and international cooperation.

Figure 4-58 Latin America Ministerial Roundtable

(d) Electrolyser manufacture is in its early stages, but growth is picking up. Production of H_2 from electrolysers is still in its early stage. Europe leads the world with a manufacture capacity of 1.2 GW per year (Figure 4-59). Such capacity is supposed power more than 500,000 passenger cars using H_2 produced from water. The capacity is growing rapidly throughout Europe.

For example, the world largest electrolyser plant, under construction by the United Kingdom's ITM Power, is expected to produce 1 GW per year. In addition, NEL Hydrogen of Norway has announced plans to build a plant with a production capacity of 360 MW annually and the potential to triple the productivity.

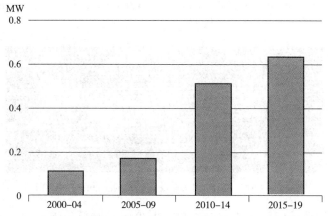

Figure 4-59 Average installed size of electrolyser projects, 2000—2019 (from IEA)

(e) The Clean Energy Ministerial Hydrogen Initiative (CEM H2I, Figure 4-60). This program has many participating countries from developed and developing countries. The 21 member countries include Australia, Austria, Brazil, Canada, Chile, China, Costa Rica, European Commission, Finland, Germany, India, Italy, Japan, The Netherlands, New Zealand, Norway, Saudi Arabia, South Africa, South Korea, United Kingdom and United States.

Figure 4-60 Hydrogen initiative logo

CEM H2I program promotes international collaboration on policy making, and developing projects to accelerate commercial implementation of H_2 and fuel cell technologies across all sectors of economies. Hydrogen Energy Ministerial Meeting in 2018 in Japan has recommended international cooperation to build on successes upon preliminary projects such as Hydrogen Challenge under Mission Innovation, the ongoing work through the International Partnership for Hydrogen and Fuel Cells in the economy and global analysis carried out through the IEA.

The following aspects will be specifically addressed: (ⅰ) Ensure successful deployment of H_2 with current industrial applications. (ⅱ) Ensure implementation of H_2 technologies in transportation sector for freight, mass transit, light-rail and marine. (ⅲ) Explore new roles of H_2 meeting demands from communities.

Members of the initiative are working on ambitious plans to develop programs to identify national targets and accelerate global aspirational goals for H_2 cities.

(f) H_2-based fuel cell systems for transportation vehicles. China has launched several strategic programs to develop high power fuel cell products for transportation purposes. HYNOVATION "KUN" series of high-power fuel cell system-KUN 132 – 001 has passed the compulsory test, with the rated power of 132 kW (Figure 4 – 61). Currently, HYNOVATION has formed 50 ~ 132 kW whole series of H_2 fuel cell system and key components products chain, which could meet the demands of commercial H_2 fuel cell vehicles and related application scenarios of power requirements.

Figure 4 – 61 Fuel cell engine with power output > 130 kW

The high-power H_2 and fuel cell engine of SinohyTech carried by Beiqi Futon 32t Hydrogen fuel cell heavy-duty truck have a rated power of 109kW, which has passed the national GB/T 24554—2009 *Performance Test Methods for Fuel Cell Engines* and normal cold start up test at – 30℃ (Figure 4 – 62). The Hydrogen FCE adopts fuel cell stacks with independent intellectual property rights.

Figure 4 – 62 Heavy trucks using fuel cells

4.11.3　Carbon-based H_2 production

(1) Steam reforming

Steam reforming is often conducted at temperatures ranging from 500℃ to 900℃ at > 2MPa pressure. Low temperature operations are feasible when oxygenates are used as feeds. Industrial applications include installation of a packed bed reactor with external heating, with bed materials mainly consisting of non-precious metal catalysts in an inert support structure.

In reforming of CH_4, Ni-based metal oxide catalysts are used as kinetic limitations and catalyst activity are insignificant limiting factors. Steam to carbon ratio and reaction temperatures are two key process parameters. Conventional plants are generally operated with steam/carbon ratio of 2.5~3 with a remarkable thermal efficiency of 85%.

Advanced reaction engineering techniques such as micro-channel reactors are applied to address existing issues in mass and heat transfer limitation in traditional reactors. Those systems also have advantages from the deployment of expensive noble metals because high specific activity with Ni and inexpensive catalysts can be realized practically.

(2) Partial oxidation

Syngas production from heavy oil and coals is achieved by burning fuels under substoichiometric O_2/fuel ratio, in the presence or absence of catalyst materials. For non-catalytic processes, operating temperature usually ranges from 1200~1500℃ at pressures of 2.5~8.0MPa to ensure complete fuel conversion and prevent formation of coke and soot. If Ni or Rh catalysts are used, the reaction temperature can be lowered by approximately 300~400℃.

(3) Autothermal reforming

Combining autothermal reforming with partial oxidation results in an almost enthalpy zero scheme. The heat required in reforming can be supplied by partial oxidation step. Reforming reactions are usually carried out at 900~1500℃ and 0.1~8.0MPa pressure. Reforming often produce additional H_2 products compared with partial oxidation, offering a flexible way by varying H/C ratio and O/C ratios in the products.

(4) Gasification

Solid fuels such as pine wood and wasted biomass can be converted to syngas products *via* gasification in O_2 and steam atmosphere. The amount of oxidant and steam can be tuned according to the composition of solid fuels as such yield of target product gas composition can be achieved. The main issue plaguing with gasification technology is the formation of tar. Biomass feedstocks often product higher amounts of tars thus demand prolonged residence time to ensure intensive conversion and product purification.

4.11.4　H_2 upgrading and purification

(1) Water gas shift reaction

Impurities in H_2 fuel may have detrimental influence on fuel performances. H_2 produced from syngas by reforming and gasification requires a shift reaction to convert CO into CO_2 and additional H_2. The process is often conducted in a two-staged adiabatic process with a high-temperature shift (HT-WGS) and low temperature shift (LT-WGS) sections. HT-WGS is usually operated at temperatures higher than 350℃. HT shift limit reaction equilibrium resulting in a CO content of 3% in effluents. LT shift favors an equilibrium towards more H_2 generation and less CO (0.1%) in outlet stream.

Feand Cu-based catalysts are often used for HT and LT reactors, respectively, although alternative catalysts can also be used with better performances.

(2) Adsorption

Impurities in H_2 feed can be adsorbed on solid materials due to different polarity and molecular interaction with surface functional groups on porous materials. Pressure and temperature swing adsorption processes are typical in practical operations. Pressure swing processes can produce H_2 purity up to 99.9999%. Regeneration of adsorbent is performed by varying pressure or temperature to release adsorbed gases.

(3) Absorption

The absorber material can be liquid solvents or solid materials. The absorption can occur as a physical or chemical interaction. Different from adsorption processes, regeneration of absorber materials is conducted at elevated temperature or pressure. The most common absorption process is removal of CO_2 *via* liquid absorbent in H_2 effluent, in a scrubber column. For chemical absorption, an aqueous amine solution is used for CO_2 removal.

(4) Chemical looping H_2

Chemical looping generated H_2 is an innovative process which utilizes oxides to convert syngas to H_2 and CO_2. Interestingly, water is used to regenerate metal oxide materials which contain lattice oxygen for oxidation of CO to CO_2. The reaction in oxidation step is similar to chemical looping. Lattice oxygen in metal oxide such as Fe, Ni, Cu, and Ce can oxidize CO into CO_2. Regeneration of metal oxide is carried out in the presence of steam instead of air. Metal species reduced during oxidation of CO can be regenerated by oxidation by water. Considering the economics and environmental impact, High melting metal oxides, such as Al_2O_3, CeO_2, and MgO, are added to increase the solid stability and improve the reactivity.

Fuel reactor: $Fe_2O_3 + CO/H_2 \longrightarrow Fe + CO_2/H_2O$

Steam reactor: $Fe + H_2O \longrightarrow Fe_3O_4 + H_2$

Air reactor: $Fe_3O_4 + O_2 \longrightarrow Fe_2O_3$

Summary

This chapter has listed and discussed mechanism, configuration and performances of various fuel cell devices. Students should understand the reaction mechanism involved in hydrogen oxidation and oxygen reduction reactions. The configuration of various types of fuel cell is important for students with engineering background. Catalyst, the key part for oxygen reduction, should be fully understood for students with science and engineering background. Other fuel cell parts, including biopolar plate, exchange membrane and sealing section are also important. Electrochemical methods can be applied to study the active sites of metal catalysts.

For other types of fuel cells, the performance curves should be compared and students should be able to explain how process parameters, including purity of fuels and oxidants affect the performance and durability of fuel cell devices.

Exercises for Chapter 4

(1) Thermodynamics for fuel cells using H_2 and O_2.

(2) Configuration of PEMFC, AFC, PAFC, MCFC, SOFC, *etc*.

(3) Power output and efficiency for fuels and application scenarios.

(4) Impact of impurities on different fuel cells.

(5) Origins of different performance losses and rationale the solution.

References and Notes

[1] Renewables 2020: Analysis and forecast to 2025. International Energy Agency, 2020 (www.iea.org).

[2] Green Chemistry: Process Technology and Sustainable Development, Tatsiana Savitskaya, Iryna Kimlenka, Yin Lu, et al. Springer (ISBN 978-981-16-3745-2, ISBN 978-981-16-3746-9 (eBook)), 2021: 15-92.

[3] Industrial Catalytic Processes for Fine and Specialty Chemicals, Sunil S. Joshi, Vivek V. Ranade. Elsevier (ISBN 978-0-12-801457-8), 2011: 41-106.

[4] Solar energy and photonics for a sustainable future. A white paper from the 7th Chemical Sciences and Society Summit (CS3), 2017 (www.chemsoc.org.cn).

[5] Metal Oxides/Chalcogenides and Composites Emerging Materials for Electrochemical Water Splitting, Aneeya Kumar Samantara, Satyajit Ratha. Springer (ISBN 978-3-030-24860-4, ISBN 978-3-030-24861-1), 2019: 31-40.

[6] Science to enable sustainable plastics. A white paper from the 8th Chemical Sciences and Society Summit (CS3), 2020 (rsc.li/progressive-plastics-report).

[7] Anisotropic Metal Chalcogenide Nanomaterials Synthesis, Assembly, and Applications, Geon Dae Moon. Springer (ISBN 978-3-030-03942-4, ISBN 978-3-030-03943-1), 2018: Chapter 1.

[8] Photoelectrochemical Water Splitting: Materials, Processes and Architectures, Hans-Joachim Lewerenz and Laurence Peter. RSC Energy and Environment Series (ISBN 978-1-84973-647-3), 2013: 52-105.

[9] Photoelectrochemical Solar Conversion Systems: Molecular and Electronic Aspects, Andrés G. Muñoz. CRC Press (ISBN-13: 978-1-4398-6926-0), 2012: 1-133.

[10] Solar Based Hydrogen Production Systems, Ibrahim Dincer, Anand S. Joshi. Springer (ISBN 978-1-4614-7430-2), 2013: 27-69.

[11] Heterogeneous Photocatalysis using Inorganic Semiconductor Solids, Umar Ibrahim Gaya. Springer (ISBN 978-94-007-7774-3), 2013: 91-135.

[12] Nanostructured Photocatalysts: Advanced Functional Materials, Hiromi Yamashita, Hexing Li. Springer (ISBN 978-3-319-26077-8), 2016: 79-99.

[13] Photoelectrochemical Solar Cells, Nurdan Demirci Sankir, Mehmet Sankir. Wiley (LCCN 2018044559), 2019: 59-145.

[14] Photochemical Water Splitting: Materials and Applications, Neelu Chouhan, Ru-Shi Liu, Jiujun Zhang. CRC Press (ISBN-13: 978-1-4822-3759-7), 2016: 41-80 + 161-209.

[15] Solar Energy Conversion: The Solar Cell (2nd ed), Richard C. Neville. Elsevier (ISBN: 0-444-89818-2), 1995: Chapter 1.

[16] Photoelectrochemical Water Splitting: Standards, Experimental Methods, and Protocols, Zhebo Chen, Huyen N. Dinh, Eric Miller. Springer (ISBN 978-1-4614-8297-0), 2013: 17-103.

[17] Photoelectrochemical Hydrogen Production, Roelvan de Krol, Michael Gratzel. Springer (ISBN 978 - 1 - 4614 - 1379 - 0), 2011: 121 - 173.

[18] Advances in Photoelectrochemical Water Splitting: Theory, Experiment and Systems Analysis, S. David Tilley, Stephan Lany, Roel van de Krol. Royal Society of Chemistry (ISBN 978 - 1 - 78262 - 925 - 2), 2018: 29 - 58.

[19] Photoelectrochemical Solar Fuel Production: From Basic Principles to Advanced Devices, Sixto Giménez, Juan Bisquert. Springer (ISBN 978 - 3 - 319 - 29639 - 5), 2016: 163 - 199.

[20] Inorganic Metal Oxide Nanocrystal Photocatalysts for Solar Fuel Generation from Water, Troy K. Townsend. Springer Theses (ISBN 978 - 3 - 319 - 05241 - 0), 2014: 27 - 37.

[21] Photocatalysis: Fundamentals, Materials and Applications, Jinlong Zhang, Baozhu Tian, Lingzhi Wang, et al. Springer (ISBN 978 - 981 - 13 - 2112 - 2), 2018: 75 - 166.

[22] Advanced Materials for Renewable Hydrogen Production, Storage and Utilization, Jianjun Liu. AvE4EvA (ISBN - 10: 953 - 51 - 2219 - 3), 2015: Chapter 1 - 2.

[23] Visible-Light-Active Photocatalysis: Nanostructured Catalyst Design, Mechanisms and Applications, Srabanti Ghosh. Wiley-VCH (ISBN 978 - 3 - 527 - 34293 - 8), 2018: 117 - 157.

[24] Sustainable Hydrogen Production Processes: Energy, Economic and Ecological Issues, José Luz Silveira. Springer (ISBN 978 - 3 - 319 - 41614 - 4), 2016: Chapter 1.

[25] Advanced Processes in Oxidation Catalysis: from Laboratory to Industry, Daniel Duprez, Fabrizio Cavani. Imperial College Press (ISBN 978 - 1 - 84816 - 750 - 6), 2014: Chapter 1.

[26] Systems Life Cycle Costing: Economic Analysis, Estimation and Management, John Vail Farr. CRC Press: Taylor & Francis Group (ISBN - 13: 978 - 1 - 4398 - 2892 - 2), 2011: 1 - 12.

[27] Product Stewardship: Life Cycle Analysis and the Environment, Kathleen Sellers. CRC Press: Taylor & Francis Group (ISBN 13: 978 - 1 - 4822 - 2330 - 9), 2015: 5 - 42.

[28] Life Cycle Assessment of Renewable Energy Sources, Anoop Singh, Deepak Pant, Stig Irving Olsen. Springer (ISBN 978 - 1 - 4471 - 5363 - 4, ISBN 978 - 1 - 4471 - 5364 - 1 (eBook)), 2013: 37 - 131.

[29] New and Future Developments in Catalysis: Catalytic Biomass Conversion, Steven L. Suib. Elsevier (ISBN 978 - 0 - 444 - 53878 - 9), 2013: 73 - 89.

[30] Production of Platform Chemicals from Sustainable Resources, Zhen Fang, Richard L. Smith, et al. Springer (ISBN 978 - 981 - 10 - 4171 - 6), 2017: Chapters 5 - 12.

[31] Fuels, Chemicals and Materials from the Oceans and Aquatic Sources, Francesca M. Kerton, Ning Yan. Wiley (LCCN 9781119117162), 2017: 151 - 180.

[32] Lignin and Lignans as Renewable Raw Materials: Chemistry, Technology and Applications, Francisco G. Calvo-Flores, Jose A. Dobado, Joaquin Isac-Garcia, et al. Wiley (ISBN 9781118597866), 2015: 289 - 308.

[33] Nanoporous Catalysts for Biomass Conversion, Fengshou Xiao, Liang Wang. Wiley (ISBN 9781119128090), 2018: 79 - 145.

[34] Martin Winter and Ralph J. Brodd, What Are Batteries, Fuel Cells, and Supercapacitors? Chemical Review [J], 2004, 10, 4245 - 4270.

[35] Non-Noble Metal Fuel Cell Catalysts, Zhongwei Chen, Jean-Pol Dodelet, Jiujun Zhang. Wiley-VCH (ISBN 978 - 3 - 527 - 33324 - 0), 2014: 183 - 204.

[36] Hydrogen and Fuel Cells: Emerging technologies and applications, Bent Sørensen. Elsevier (ISBN 0 – 12 – 655281 – 9), 2005: 1 – 88.

[37] Fuel Cell Science and Engineering: Materials, Processes, Systems and Technology Vol 1, Detlef Stolten, Bernd Emonts. Viley-VCH (ISBN 978 – 3 – 527 – 33012 – 6), 2012: 67 – 126.

[38] Fuel Cell Handbook (7th ed.), EG&G Technical Services, Inc. 2004: Chapters 1 – 7.

[39] PEM Fuel Cell Electrocatalysts and Catalyst Layers: Fundamentals and Applications, Jiujun Zhang. Springer (ISBN 978 – 1 – 84800 – 935 – 6), 2008: 89 – 129.

[40] Chemical Energy Storage, Robert Schlogl. De Gruyter (ISBN 978 – 3 – 11 – 026407 – 4), 2013: 87 – 124.

[41] Handbook of Electrochemical Energy, Breitkopf Swider-Lyons. Springer (ISBN 978 – 3 – 662 – 46656 – 8), 2017: 591 – 641.